Low Power FM

by Sharon M. Scott

Low Power FM For Dummies®

Published by: **John Wiley & Sons, Inc.,** 111 River Street, Hoboken, NJ 07030-5774, www.wiley.com

For general information on our other products and services, please contact our Customer Care Department within the U.S. at 877-762-2974, outside the U.S. at 317-572-3993, or fax 317-572-4002. For technical support, please visit https://hub.wiley.com/community/support/dummies.

Wiley publishes in a variety of print and electronic formats and by print-on-demand. Some material included with standard print versions of this book may not be included in e-books or in print-on-demand. If this book refers to media such as a CD or DVD that is not included in the version you purchased, you may download this material at http://booksupport.wiley.com. For more information about Wiley products, visit www.wiley.com.

Library of Congress Control Number: 2023945274

ISBN 978-1-394-18532-0 (pbk); ISBN 978-1-394-18533-7 (ebk); ISBN 978-1-394-18534-4 (ebk)

SKY10054865_091223

Table of Contents

Introduction 1

Part 1: Lighting the Fuse: Getting High Impact from Low Power 5

CHAPTER 1: Getting High Impact from Low Power 7

CHAPTER 2: Finding or Founding Your Nonprofit 19

CHAPTER 3: Getting Permission to Broadcast 39

Part 2: Countdown to Blastoff: Launching Your Broadcast 59

CHAPTER 4: Firing Up Your Frequency 61

CHAPTER 5: Weaving Your Webcast 75

Part 3: Staying in Orbit: Building a Station That Endures 89

CHAPTER 6: Where the Magic Happens: Building a Studio with a Spark! 91

CHAPTER 7: Cranking It Up: Developing Your Music Collection 111

CHAPTER 8: Going for the Gold: Developing Station Income 127

CHAPTER 9: Holding the Line: Managing Station Expenses 141

Part 4: Star Systems: (Low) Power to the People! 157

CHAPTER 10: Keeping It Legit: Broadcasting Rules and Responsibilities 159

CHAPTER 11: It's Showtime! Developing Programming and Taking It Live 175

CHAPTER 12: Forming the Band: Building Your Crew 197

Part 5: Riding the Waves: Ups and Downs in Community Broadcasting 209

CHAPTER 13: Understanding Murphy's Law of Radio 211

CHAPTER 14: Cultivating Community Connections 225

CHAPTER 15: Connecting with Listeners Locally and Worldwide 241

Part 6: The Part of Tens 255
CHAPTER 16: Ten Ways to Fill Dead Air (and Come Alive!) 257
CHAPTER 17: Ten Epic Radio Broadcasts 265
CHAPTER 18: Ten Inspiring Insights from Low Power FM Station Founders 275

Appendix A: Directory of Low Power FM Radio Stations 285

Appendix B: Glossary 293

Index 299

Table of Contents

INTRODUCTION 1
About This Book 1
Foolish Assumptions 2
Icons Used in This Book 3
Beyond the Book 4
Where to Go from Here 4

PART 1: LIGHTING THE FUSE: GETTING HIGH IMPACT FROM LOW POWER 5

CHAPTER 1: **Getting High Impact from Low Power** 7
Getting to Know Low Power FM Radio 8
Understanding the American radio landscape 9
Sailing on the open seas 9
Switching on the LPFM light 11
Introducing the Local Community Radio Act 12
Envisioning Your Radio Station 13
Why radio? 14
Why now? 15
Why you? 15
Creating Your FM Dreamscape 16

CHAPTER 2: **Finding or Founding Your Nonprofit** 19
Preparing for the LPFM Application Window 20
Meeting LCRA expectations 20
Keeping it local 20
Upholding board expectations 20
Scoring Priority Points 22
Uniting for Good: Partnering with an Existing Nonprofit 23
Starting from Scratch: Forming a New Nonprofit 24
Appointing your board of directors 24
Naming your organization 27
Defining your Mission 27
Focusing your vision 28
Incorporating your nonprofit 29
Drafting your by-laws 30
Applying for your Employer Identification Number 30
Establishing and maintaining 501(c)3 status 31
Getting a state sales tax exemption 32
Filing 990s 32
Obtaining nonprofit postage 33

Protecting yourself and your assets 33
Making connections and receiving funds. 34
Making Your Mark: Branding Your Organization 34
Developing a logo 35
Building a website 36
Establishing a presence on social media 36
Building community connections 36

CHAPTER 3: **Getting Permission to Broadcast** 39
Finding Your Frequency 39
Locating quiet spots on the dial 40
Securing an engineering study 41
Hunting for Your Tower 42
Taking the first steps 43
Negotiating with a landlord 45
Applying for an FM Frequency 46
Registering with the FCC 46
Collecting the necessary materials 47
Submitting your application 49
Winning Your FM Frequency 50
Navigating MX Situations 51
Winning your frequency 51
Joining a timeshare 54
Alternative Methods for Getting on the FM Dial 55
Working with licensees in your community 55
Using Part 15 broadcasting 56

PART 2: COUNTDOWN TO BLASTOFF: LAUNCHING YOUR BROADCAST 59

CHAPTER 4: **Firing Up Your Frequency** 61
Permission Granted: Getting Started 61
Keeping your FCC information safe 62
Claiming your call sign 62
Setting Up Your Signal 63
Sourcing equipment housing 63
Piecing everything together 64
Choosing Between On-Site and Remote-Site Transmission 65
Understanding Your FM Transmitter 66
Installing Your Antenna 68
Installing the Emergency Alert System 70
Lighting It Up! 71
Keeping up with the FCC 72
Renewing your license 72
Keeping the light on 73
Partying at Your Official FM Launch 74

CHAPTER 5: **Weaving Your Webcast** 75
Launching Your Station on the Infinite Dial 78
Connecting with a stream host 79
Getting a direct URL 79
Embedding player buttons on your website 80
Utilizing pop-up players 81
Adding the audio element 81
Registering with Online Radio Directories 84
Developing Apps for Listeners on The Go 85
Teaching Skills to Smart Speakers: "Alexa, Play WXOX" 88

PART 3: STAYING IN ORBIT: BUILDING A STATION THAT ENDURES 89

CHAPTER 6: **Where the Magic Happens: Building a Studio with a Spark!** 91
Finding a Studio Location 92
Considering Interior Spaces 94
Soundproofing 95
Sound dampening 95
Installing Studio Furniture 95
Selecting Studio Equipment 97
Audio console 98
Microphones 99
Mic booms 100
On-air light 100
Headphones 101
Headphone amp 101
Studio monitors 102
Turntables 102
CD players 104
Cassette players 105
Computers 105
Computer cables 106
Accessories and adapters 107
Providing Community Access and Security 108

CHAPTER 7: **Cranking It Up: Developing Your Music Collection** 111
Building a Physical Library 112
Thinking about shelving (and space) 113
Choosing an organizational method 115
Collecting your materials and getting to work 116
Venturing into vinyl 116
Learning CD selectivity 117

Tangling with tapes 118
Programming albums 118
Marking materials for safekeeping 120
Developing a Digital Library 121
Seeking digital selections 121
Managing digital files 122
Using software to keep the music going 123
Archiving Original In-House Productions 125

CHAPTER 8: **Going for the Gold: Developing Station Income** 127
Making Friends and Creating Opportunities 128
Organizing a Capital Campaign 128
Building Station Membership 131
Developing Partnerships and Sponsorships 133
Working with Underwriters 134
Finding Government Money 136
Considering Incentive Programs 136
Driving Vehicle Donations 137
Maximizing Merch Sales 138
Hosting Fundraising Events 138

CHAPTER 9: **Holding the Line: Managing Station Expenses** 141
Maintaining Your Home-Base 142
Factoring in The Price of Technology 143
Internet service 143
Webhosting 143
Remote access 143
Video conferencing 144
Licensing Music 144
FM transmission 144
Digital transmission 145
Hiring Personnel and Growing a Happy Team 147
Showing staff and donor appreciation 148
Funding continuing education 149
Joining Professional Organizations 150
Budgeting for Communications 151
Keeping Financial Records 152
Bookkeeping 152
Filing annual taxes 153
Maintaining transparency 153
Remembering Everything Else, Too 154
Saving up for sundries 154
Buying merch 154
Launching events 155
Saving for station updates 156

PART 4: STAR SYSTEMS: (LOW) POWER TO THE PEOPLE! 157

CHAPTER 10: Keeping It Legit: Broadcasting Rules and Responsibilities 159
Seeking Help and Getting Legal Advice 160
Understanding FCC Laws 160
Delving into The Public and Broadcasting 162
Remembering First Amendment rights 163
Abiding by FCC Laws 163
Abiding by basic laws and policies 164
Avoiding hoaxes and distorted reporting 164
Balancing political broadcasts 165
Watching out for objectionable programming 165
Following other content regulations 167
Staying noncommercial 169
Fundraising in the right way 170
Announcing program-related materials 170
Steering clear of payola 171
Moving Beyond the FCC: Other Legal Considerations 171
Avoiding defamation of character 171
Understanding music licensing 172
Deciding on your house rules 173

CHAPTER 11: It's Showtime! Developing Programming and Taking It Live 175
Moving from Dreams to the FM Dial 176
Music programs 177
Spoken word programs 178
Experimental radio programs 179
Setting the Scene 180
Deciding on program titles and DJ names 180
Making musical selections 181
Considering guests for your show 182
Promoting your radio program 183
Preparing Your Shows 184
Checking equipment 185
Cueing music 185
Going Live! 186
Being "in the now" 186
Setting the scene 187
Handling surprise situations 187

Keeping Time with the Programming Elements 187
Announcing legal identifiers 189
Reading the news 189
Making public service announcements 191
Reading out promos 192
Detailing an events calendar 193
Hosting On-Air Guests 193
Minding Studio Etiquette 194

CHAPTER 12: **Forming the Band: Building Your Crew** 197
Developing the Station CORE 198
Recruiting and Training Your On-Air Team 200
Thinking about the schedule 201
Handling applications 201
Vetting new recruits 202
Overseeing orientation 202
Putting your recruits through basic training 202
Addressing Staff Responsibilities 204
Covering scheduled shifts 204
Attending staff meetings 206
Representing the station on social media 207
Cultivating a respectful workplace 207
Suspension of shows 208

PART 5: RIDING THE WAVES: UPS AND DOWNS IN COMMUNITY BROADCASTING 209

CHAPTER 13: **Understanding Murphy's Law of Radio** 211
Being Prepared When the Broadcast Goes Silent 212
User error 213
Utility failure 213
Radio silence 214
Streaming silence 214
Dealing with Static and Other Interference 215
FM interference 216
Stream glitching 217
Disappearing (or Delayed) Talent 218
Failing Equipment 218
Spilling onto the Soundboard 219
Breaking the Rules 219
Staying Tough and Carrying On 221

CHAPTER 14: **Cultivating Community Connections** 225
Bringing the Community to the Station 226
Hosting open studio events 226
Interviewing great guests 227
Reporting on community trauma 230
Bringing power to the people! 230
Working with youth groups 231
Telling veterans' stories 231
Organizing DJ training programs 232
Bringing the Station to the Community 234
Taking part in parades and festivals 234
Hosting station spectacles 236
Covering local events 236
Reporting on Emergencies 238

CHAPTER 15: **Connecting with Listeners Locally and Worldwide** 241
Promoting Your Station Internationally 242
Getting your swag on 242
Sewing up the socials 242
Getting published 243
Connecting at conferences 243
Syndicating Programs Worldwide 245
Pacifica Affiliate Network 245
Public Radio eXchange 245
Providing On-Demand Listening 248
Ark and Confessor 249
Mixcloud 249
Soundcloud 249
Stitcher 250
Rambling Remotes 250
Establishing the DJ stream 250
Setting up audio inputs 251
Taking flight 252
Entering a brave new world! 253

PART 6: THE PART OF TENS 255

CHAPTER 16: **Ten Ways to Fill Dead Air (and Come Alive!)** 257
Explore Your Music Stacks 257
Utilize Copyright-Free Material 258
Amplify Your Voice 259
Conduct an Interview 260

Play an Instrument 260
Rally Local Performers 261
Simulcast Another Low Power FM Station....................................261
Place a Mic Outside Your Window 262
Find a Numbers Station 262
Get Experimental....................................263

CHAPTER 17: **Ten Epic Radio Broadcasts** 265
The Birth of Radio (March 1, 1893)....................................265
The World's First DJ (April 1, 1914) 266
The First Live Broadcast of Presidential Election Results (November 2, 1920)....................................267
The Broadcast of War of the Worlds (October 30, 1938)....................................267
The Launch of KPFA and the Pacifica Network (April 15, 1949) 268
The Launch of First Black Owned and Operated Station in the U.S. (October 3, 1949) 269
Bob Fass's Fly-In (February 11, 1967)....................................270
The Launch of Radio Free Alcatraz (December 22, 1969) 271
The Reporting of Hurricane Katrina and Dome City Radio (August 2005)....................................272
The Power of International Human Rights Simulcasts (2020 and Beyond) 273

CHAPTER 18: **Ten Inspiring Insights from Low Power FM Station Founders**....................................275
LaGanzie Kale, KLEK Jonesboro....................................275
Julianna Ross, KMGP Seattle 276
Shawn Campbell, WCXP Chicago....................................277
Max Kelley, WAYO Rochester....................................279
Clark Grant, KBMF Butte....................................279
Davyne Dial, WPVM Asheville....................................280
Kara Richardson, WXND Louisville 281
David Beaton, WSLR Sarasota 282
Reverend Joan C. Ross, WNUC Detroit....................................282
Stella Linder Byrne, KMRD Madrid 283

APPENDIX A: DIRECTORY OF LOW POWER FM RADIO STATIONS 285

APPENDIX B: GLOSSARY 293

INDEX....................................299

Introduction

The airwaves belong to you. You breathe them in; you breathe them out. Yet, they endlessly elude your grasp. Radio is everywhere all at once, yet it is nowhere at all. A shape-shifter and a time traveler, so much can be done with this ever-mysterious medium, and we've only just begun to experiment with its potential.

The Low Power FM movement is a small but energetic alliance. Our army is growing by the day. Real living, breathing humans are pushing through the cracks of commercial automation with spontaneous ideas and original thought. DJs are playing music because it moves them, not because it fits a mathematical formula. Listeners are discovering these new vibrant stations, where real DJs speak to them in real time, about real things that are going on. They find that no algorithm can simulate the effect of connecting with another human. Radio penetrates the mind like no other medium! It works because it inspires the listener's imagination and allows them to be a part of the creation. This book is your backstage pass, your entrance behind the curtain!

About This Book

Low Power FM For Dummies provides you with the information you need to start and maintain a noncommercial FM radio station — to create a spot on the terrestrial dial where you can amplify the voices of your community, play the music they love, and address the issues that are important to them. It begins with forming your nonprofit organization and walks you through all the steps of building your own community media outlet. This includes assembling your equipment, training your DJs, launching your web stream, and sharing your programming worldwide.

Starting with the basics, this book walks you through the steps of building a new community radio station. I share with you experience I have learned and the resources I have gathered on my own journey in the hope that it will shine a light on yours. In cases where you may need further expertise, I point you in the right direction. The suggestions in this book may not always work for your situation and you may find other solutions that you feel work better instead. Fantastic! Write in the margins, reach out and let me know, and maybe we can spread the word or get it in the next edition. This is a living, growing community project after all! My greatest hope is that this book gives you the confidence to find your own direction in radio — to help your station reflect the unique personality of your community and become unlike anything else that has ever hit the air! To make the content of this book more accessible, I've divided it into six parts:

- **Part 1: Lighting the Fuse: Getting High Impact from Low Power.** This part helps you understand what it takes to become a Low Power FM (LPFM) broadcaster and shows you how to get there. It begins with the background of the LPFM movement and ends with you getting permission to broadcast.
- **Part 2: Countdown to Blastoff: Launching Your Broadcast.** Here you discover the logistics for getting your signal on the FM dial and streaming online around the world.
- **Part 3: Staying in Orbit: Building a Station that Endures.** Beyond the broadcast itself, there is a lot to think about when developing a new radio station. This part discusses everything else, from assembling your studio to developing your music collection, as well as the ins and outs for keeping it all afloat.
- **Part 4: Star Systems: (Low) Power to the People!** Nothing is more valuable to a station than the people that keep it going. This part helps you develop guidelines for on-boarding DJs and building your backstage crew. It leads you through on-air regulations and helps you develop your own station policies.
- **Part 5: Riding the Waves: Ups and Downs in Community Broadcasting.** So many adventures are experienced in broadcasting, so it's hard to contain them all in a single category. This part includes those unruly ups and downs that you may encounter when running a radio station.
- **Part 6: The Part of Tens.** The book concludes with a fun section that you can refer to when you need a breather, a brainstorm, or a piece of friendly advice. Radio does not always have to be so serious!

Foolish Assumptions

This book is written for people with the desire to start and lead their own noncommercial FM or online radio stations, without any prior knowledge or experience in the field. However, the information within is also helpful to anyone involved with a nonprofit community radio station — board members, DJs, technical directors, and even superfans who just want to learn more about how the whole thing works.

The Low Power FM movement was born out of the *Local Community Radio Act* as a means for giving local nonprofit organizations access to the airwaves. My assumption here is that most charities are busy with other activities and have little or no broadcasting experience. Luckily, radio does not require a great deal of technical knowledge, especially after you get going. But you do need a desire to communicate and a willingness to learn. Although this book focuses on building LPFM radio stations, it will also be helpful to those interested in launching community stations online. Most of the principles are the same except that you have less regulation online.

The term *community station* refers specifically to noncommercial stations operated by members of the immediate community. Community stations, for the most part, are powered by volunteer DJs, supported by listener donations, and broadcast local productions. I am the co-founder of ART FM, the nonprofit organization that manages WXOX 97.1 FM in Louisville, Kentucky. Throughout the text I refer to the station as both ART FM and WXOX. At times, ART FM is used to refer to the organization itself and the online broadcast with WXOX being more specific to the FM transmission.

As you read, please keep in mind that I am a LPFM radio station manager. I am not an attorney or an engineer. Neither am I a radio network representative or an agent of the Federal Communications Commission. I am a person, just like you, but with 10 years of LPFM experience. Many people have reached out over time, asking for advice. In addition to WXOX, I've helped a few other stations (such as WXNA and WXND) get on the air. In this book, I write it all down in the most straightforward way I know how. If in doubt, always consult with a professional (be that an attorney, an engineer, an electrician, or a pastry chef!).

Icons Used in This Book

Acting as traffic conductors for this book, the little icons you spot in the page margins tell you when you need to slow down and when you can fast forward. Here's the key to understanding what they mean.

REMEMBER

Holding a LPFM broadcasting license is a serious business! This icon alerts you to key information that's important for keeping your station on air and acting for the public benefit. Pay attention at the back!

TECHNICAL STUFF

This icon lets you know when the waters are getting deep. If you feel like this material is getting over your head, skip over it or circle back to it later (after you have more of the basic concepts mastered).

TIP

When you see this icon, you're getting the benefit of my wisdom — experience and helpful advice I've picked up along the way to make your own journey easier.

WARNING

When you spot this icon, be sure to read it! Snakes, Black Holes, and quicksand await you on your LPFM journey, but so do stars, lighthouses, and song larks! Whatever you do, pay attention to this information as it will help you navigate your ship safely through LPFM's dangerous waters.

Beyond the Book

In addition to the advice and guidance provided in this book, you get online access to even more help and information about building a Low Power FM radio station. I've included a number of important resources to help you manage the business of running your LPFM station — you can access them here: `www.dummies.com/go/lowpowerfmfd`.

As well as these documents, you can find this book's handy online Cheat Sheet by heading to `www.dummies.com` and searching for "Low Power FM For Dummies Cheat Sheet."

Where to Go from Here

Consider this book a user guide to flip backwards and forwards through, rather than any kind of linear story. You can skip through sections of text and focus on the portions that are helpful to you. The engineering and legal details within this book apply specifically to Low Power FM radio stations operating in the United States. However, the majority of this book is relevant to noncommercial stations operating anywhere in the world — full power or micro-broadcaster, FM or online.

Stream one of the LPFM stations listed in the directory at the back of the book. Use the index and the table of contents to dive directly into the subjects that most interest you. If you want to know where the LPFM movement came from, go ahead and start with Chapter 1. If you want to jump directly into starting a station online, skip to Chapter 5. If your community station is already on air and you're seeking some station management advice, dive into Part 3. If you want a taste of what it's like to run a LPFM before taking the leap, Chapter 18 provides helpful insights from LPFM station founders. It's OK to start with dessert!

1 Lighting the Fuse: Getting High Impact from Low Power

IN THIS PART . . .

Dig into the American broadcasting landscape and discover how you can become a part of it.

Build or join a nonprofit organization to help get your Low Power FM radio station up and running.

Get permission to broadcast and get set to take it live!

» Starting your own station

» Building on your dreams

Chapter 1
Getting High Impact from Low Power

Your radio is hungry. It craves life, music, and conversation. It longs to connect neighbors with one another and unite enemies with a song. Radio dreams of broadcasting the sound of community to the people who live there. At long last, a group of radio activists have emerged to energize and empower the FM airwaves. They have started a Low Power Revolution and they are seeking a few good recruits!

Radio is dangerous, it is live, raw, and pervasive. It is powerful and unstoppable, which may be why it became so inaccessible. For nearly a century, commercial interests have tied up the FM frequencies leaving little room in between for community voices. But not anymore! Thanks to an unlikely partnership between pirates and politicians, the *Local Community Radio Act of 2010* opened up these in-between spaces and put the power of radio into the hands of the people. If you've ever believed that you could change the world, you now have the opportunity. Take a deep breath and count to three, because an exciting journey lies ahead.

This chapter is your invitation to join the radio revolution and transform the FM dial one Low Power station at a time. Together we can bring real human voices back to the airwaves and discover new uses for this venerable medium. This chapter provides an introduction to the slow, complicated birth of the Low Power FM movement; looks at the rare, magical opportunity it has created; and considers the reasons why we must seize this chance to transform American media, one Low Power FM station at a time!

Getting to Know Low Power FM Radio

REMEMBER

Low Power FM refers to a specific class of radio stations on the American FM dial. They are licensed to nonprofit organizations to operate with 100 watts of power at 100 feet of Height Above Average Terrain or HAAT (a mathematical formula you find out more about in Chapter 4). Commonly referred to as LPFMs (as you discover throughout the pages of this book), these hyper-local broadcasters are often listener supported and volunteer driven. LPFM stations have sprung up across the country in all shapes and sizes, bringing a fresh, homegrown flavor back to America's FM dial.

Low Power FM stations have a strict limit on their broadcast power, although with the right location, 100 watts can reach thousands of people on a daily basis. Chapters 3 and 4 discuss developing an installation plan that will cover your community with the best signal possible. Just like all other stations on the FM dial, Low Power stations are regulated by the Federal Communications Commission (FCC) and have a code of law that they are required to uphold (explained in Chapter 10). This differs from those stations that stream online only; they are not required to have any kind of license or uphold FCC broadcasting expectations.

Most LPFM stations simulcast their programs via a web stream so that listeners all over the world have the chance to tune in online. Chapter 5 describes the easy steps for making this a reality. If you don't have a LPFM license, you will find that there are still many ways to amplify your voice. You can then move forward and take advantage of the rest of the information in this book about managing and maintaining a radio station. With the exception of Chapters 3, 4, and 10, the rest of the information in this book is applicable to online broadcasters.

THE MAGIC OF RADIO

The magic of radio is impossible to explain. After just one taste, you will want it again. Forever. For the rest of your life. You will sacrifice so much time and energy making it sound good, and keeping it moving forward. But radio will reward you — a million-fold as a lover and a lifesaver, a storyteller and a first responder. Radio is your trusted companion, old friend, and life on the other end of the line. Radio is a brilliant educator, a stunning entertainer, and a powerful activist. To find out more about how radio can help build communities, check out Chapters 12, 14, and 15.

Radio can be used to share music, exchange ideas, and plan revolutions. In fact, few things cannot be accomplished by speaking to thousands of people at once. Which is why radio is also so dangerous, and why regulation has previously been so strict.

Understanding the American radio landscape

There was a time — a very brief moment — when radio could have become so much more than it is today. Nikola Tesla began demonstrating the power of his new wireless invention at the dawn of the 20th century (see Chapter 17), and it would be years before anyone thought to regulate the airwaves. Many folks were experimenting with the science just to see what it could do; radio clubs, technological schools, and Boy Scouts across the country were setting up radio stations in their attics and classrooms with transmitters made from copper wire and cardboard tubes. These amateur broadcasters began friendly competitions with one another to see who could get the farthest reception, who could have the clearest sound, and so on. They also worked together, self-regulating the airwaves, scheduling airtime through local networks so that they would not interfere with one another's signals. It is hard to imagine what radio sounded like at that time. Or what might have happened had this free access of the airwaves been allowed to thrive.

Instead, federal legislation beginning with the *Radio Act of 1914* strongly favored broadcasters who had learned to monetize the airwaves, fatten their wallets, and hire the best lobbyists. As radio clubs tinkered away, these capitalist networks pushed for stringent broadcasting laws that could protect their source of income and prevent interference during their sponsorship announcements. It wasn't until the *Communications Act of 1934* and the development of a government agency to police the radio waves that the new licensing rules were taken very seriously, but this time it was drastic. Under Herbert Hoover's direction, the radio spectrum was reallocated. The largest, clearest licenses were handed out to commercial networks. Educational stations operated by clubs, universities, and nonprofit organizations were corralled into a limited area of the spectrum. The complaints of listeners, broadcasters, and educators were met with federal agents assigned to enforce the new rules. In a matter of years, commercialization had conquered the airwaves. Advertisers invented schedules that they felt sold the most product and broadcasters began sculpting their programs around the latest sales theories. To this day, mainstream American radio broadcasters cater their programming to appeal to advertising dollars, not to listener interest.

Sailing on the open seas

Low Power FM was born out of the salty waters of the *pirate radio* movement. These countless unknown individuals have sailed the dark seas of the terrestrial dial for over 100 years, insisting on the constitutional right to broadcast with or without government permission. Forced into the shadows by threats of fines, property seizure, and federal imprisonment, these independent broadcasters may only be remembered in the minds of those who experienced their fleeting moments on air. Their role in American history has vanished into the ether from which it emerged. Their work, however, broke through the barricades of federal regulation and finally made room for community voices on the FM dial.

Operating a FM transmitter without a license is illegal in the United States.

Despite this, you might be surprised at how many unlicensed broadcasters you find in your area, on air, right now. Often belonging to churches or hobbyists, these micro stations rarely mean any harm. However, radio broadcasting can have a powerful effect on the community — both physically and mentally. For example, an untrained broadcaster providing inaccurate information can cause listener panic and put lives at risk (see Chapter 10 for federal laws on responsible radio reporting), and a broadcast antenna hung too close to an occupied space can create a radiation health hazard (Chapter 4 provides advice on installing your equipment safely). These regulations make sense. The counter-argument has been that FCC laws restrict the airwaves so much that they prevent the development of hyper-local radio, including rapid emergency response. Therefore, pirates argue that they have the potential to save lives with their broadcasts and can provide more local value than the national network beamed in from some distant location.

In 1994 a pirate radio station named Free Radio Berkeley (`freeradio.org`), was caught by the FCC and fined $20,000. Stephen Dunifer, who ran the station, refused to pay the fine and vowed to fight the judgment. Dunifer's actions were inspired by Mbanna Kantako of Human Rights Radio (`humanrightsradio.net`), who had previously been cited by the FCC for pirate broadcasting but refused to pay the fine — asserting that the First Amendment protected his right to free speech.

In defiance of the FCC raid on Free Radio Berkeley, Stephen Dunifer encouraged unlicensed broadcasters nationwide to take to the airwaves in protest of restrictive federal broadcast laws that made no room for community voices. He was represented in court by the National Lawyer Guild's Committee on Democratic Communications, in the case "U.S.A. vs. Stephen Paul Dunifer." The defense focused on the First Amendment and its insistence that the government agencies use the least restrictive means possible when governing the constitutionally protected Freedom of Speech. The Guild argued that the FCC was overstepping its bounds. U.S. District Judge Claudia Ann Wilkin found their case compelling and placed a hold on the fine as she reviewed the legality of the FCC licensing system, leaving the FCC virtually powerless against renegade broadcasters. An estimated 1,000 unlicensed stations went on air during the four-year review period that followed. When the judgment finally came in against Free Radio Berkeley on a technicality, the authority of the FCC was legally confirmed by the court. Pirate radio remained illegal, and enforcement resumed in full.

Switching on the LPFM light

From its home base in West Philadelphia, however, a pirate radio station called Radio Mutiny kept pushing full steam ahead, insisting on the people's right to access the airwaves. When the FCC came knocking on the station door in 1998, the Radio Mutiny Collective responded with a savvy media campaign saying: "Community Radio — It Makes Common Sense." The press showed up and it was the beginning of a full-court press on the Federal Communications Commission.

The pirate radio community knew that there was plenty of space for small stations in between the large commercial stations on America's FM dial. Radio Mutiny teamed up with other unlicensed broadcasters to host a Pirate Radio conference in Washington, D.C., in 1998. The conference culminated in a Pirate March on Washington which caught the attention of FCC Chairman William Kennard.

As the first African American to lead the agency, Kennard hoped to rectify the loss of minority-owned stations that had occurred with the passage of the *Telecommunications Act of 1996*. When the Pirate March came past his window, he saw an opportunity. Within months, the FCC opened rule-making to develop a system for licensing community organizations broadcasting on Low Power FM stations. The order was completed and passed in just under two years.

Satisfied with this government action, Radio Mutiny spokesperson Pete Tridish announced that they were going legit. The Mutiny collective retooled and relaunched as the Prometheus Radio Project (`prometheusradio.org`), a nonprofit organization with the mission of helping nonprofits acquire FCC licenses and get on air nationwide. They acquired grants, hired employees, and traveled across the country hosting workshops while launching new stations.

Just as soon as things got moving, a series of legal attacks were launched against the LPFM movement. Among these was the *Radio Preservation Act*. This legislation, introduced by the National Association of Broadcasters (NAB) and National Public Radio (NPR) and passed by Congress in 2000, used exaggerated interference concerns to tie up urban airspace and prevent the construction of Low Power stations within heavily populated areas. Prometheus pushed forward with building stations throughout communities in rural areas while simultaneously lobbying Congress to pass another bill — the *Local Community Radio Act* (LCRA) — that would open more frequencies for LPFM stations. The collective took many creative approaches to bringing attention to community broadcasting, including assembling a hodge-podge of portable radios in front of the U.S. Capitol (shown in Figure 1-1) — perfectly symbolizing the array of unique radio stations that would take to their airwaves after the LCRA was passed. Ultimately, the new bill took 10 years, 20 attempts, and 50 handmade hula hoops to finally make way for legal community broadcasting in the United States. In late 2010, the LCRA finally got the support needed to pass Congress and head to the president's desk.

FIGURE 1-1: A number of creative demonstrations organized by the Prometheus Radio Project led to the passage of the *Local Community Radio Act*, opening spaces on the FM dial to community broadcasters.

© Pete Tridish.

Introducing the Local Community Radio Act

REMEMBER

On January 4, 2011 — thanks to a decade of tireless work by organizations such as the Prometheus Radio Project, Georgetown Law, the Future of Music Coalition, REC Networks, Free Press, the National Hispanic Media Coalition, and others — the *Local Community Radio Act* was finally signed into law by President Obama. As a result, thousands of LPFM radio stations would take to the air across the country.

But this great accomplishment did not happen without its sacrifices. Within the text of the LCRA is specific language that makes room for Low Power community broadcasters on the American dial so long as certain conditions are met. Among the stipulations is "no pirates allowed"; the law states that no one who has engaged in unlicensed radio may serve on the board of a Low Power FM station (find out more in Chapters 2 and 3). So although the pioneers — Stephen Dunifer, Pete Tridish, and Mbanna Kantako — were officially, purposefully, excluded from the new Low Power legislation, their sacrifice cracked opened the American airspace and made room for community broadcasters nationwide. From this point on, no longer would community members be excluded from the terrestrial airwaves! No longer would they be forced to listen to programming dictated by corporate sponsors. Now they had choices! At long last, American citizens became free to sculpt the airwaves to reflect their personalities and use them to amplify their concerns — but only those that seize the opportunity! And that's what this book is all about.

REMEMBER

Thanks to those early years of activism, thousands of LPFM radio stations are now on air across the United States. They come in all varieties, with studios of all shapes and sizes. Some are on university campuses, some are in community centers, some are in churches, and still others are in the basement of the station founders' homes. You do what you have to do in community radio and often you have to improvise. Somehow, it always seems worth the effort. My hope is that the number of Low Power stations will increase dramatically after the FCC opens the next application window (described in detail in Chapter 3). With any luck, your station will be one of those.

TIP

Many organizations and individuals are mentioned in this book who are willing to help. As a whole, the LPFM community is very supportive. Don't be afraid to call on them when you have questions. They understand that working with the FCC can be a complicated process. They understand that you don't have a lot of radio experience. They also understand how important it is to get local voices on the air, and they are committed to making it happen. If you want to do this you can. I believe in you! Take it step by step and one day at a time and suddenly you will be there, driving around, listening to your own community station on the FM dial!

Envisioning Your Radio Station

Radio can be everything. And more. It morphs with each moment and can turn on a dime. Outer space to underwater — I'll meet you there in an instant! It's easy to travel through sound. Yet, today, where there should be communication highways, there are black holes. Empty spaces of static and chaos that jolt the senses as you scroll across the FM dial.

Tune in the static and listen for a moment. Then, imagine this cacophony of noise slowly fading out and the voices of your friends and neighbors slowly fading in. Imagine they begin playing your favorite songs, dedicating them directly to you; soon they are talking about issues that affect your life directly, things that are happening in all the places you know. Would you ever turn to the dial again? Maybe not. Perhaps this is why corporate media resisted community radio for so long. The vibrancy of the people's broadcast can never be emulated by broadcasters who are doing it for the money.

Under the LCRA, small spots on the radio dial are pried open to make room for new 100-watt LPFM stations licensed to area nonprofits. Since its passage, nearly

2,000 LPFM stations have taken flight nationwide. Beaming with character and overflowing with personality, what these stations lack in electrical wattage, they make up for in human energy.

If you feel a little hesitant about applying for a LPFM license, I assure you that can get it done! It will be a challenge, but truly, you will get so much out of the experience that it is worth every bit of your time. Just take it on your own terms, keep it as basic as necessary as you get started, and build it up as you move along. It might not be perfect at first but just remember, anything sounds better than static, well, almost! (Discover critical things to avoid in Chapter 10.) To get started with your license application, pay attention to Chapters 2 and 3.

TIP

Few things are more rewarding than planting a garden and watching it grow. You can have a concept in mind when you start out, but it is impossible to predict exactly what your LPFM station might look like, how it might sound, or all the ways it will affect your listeners. There is no way to measure how many ears it will reach, minds it will change, hearts it will stir, or lives it will save. However, a community station *will* do all of these things and you *will* feel it happening. Flip to Chapter 18 to read the inspirational stories of ten LPFM station founders.

Why radio?

You have dozens of radio stations on your FM dial but you might think they were all the same. As you turn the dial all you find is commercials . . . and songs that lead into commercials. A few exceptions exist: sports stations, religious stations, and news stations broadcasting programming almost entirely piped in from distant locations. It's increasingly rare to find a live human voice on the radio, maybe coming from a college or high school station. When you do, hold on it! Support it! Let them know that you are out there listening.

Imagine a different life . . . waking up each morning, excited to turn on your radio, tune into any one of your favorite stations, and look forward to what surprises you might hear. Perhaps you want something chill and ambient to ease into the day. Perhaps you need something rowdy to get you out of bed and off on assignment. If you like waking up to the news, of course, you can find that, too. But what if you want an astronomy lesson? Yes! A history discussion? Yes, please! Philosophical debate? Sure, radio can do that, too! Everything is right there, just with the turn of a dial. No service plan, no pop-ups, no data tracking, just pure uninterrupted sounds enriching and educating your community. If radio sounded like this, we would be lifetimes ahead. For more programming ideas, check out Chapter 11.

Some people ask whether FM radio is dying, or if it's an outdated medium. We've been hearing these arguments since the dawn of TV, and now with the advent of the Internet, it started up all over again. Radio is a survivor. Streaming audio now allows stations to be heard all over the world (find out more in Chapters 5 and 15) and it is wonderful! Podcasting allows programs to be heard at the listener's convenience. And none of this undercuts the validity of FM radio. Quite the opposite, in fact, as an FM broadcast increases the value and accessibility of an online broadcast. It provides the credentials necessary for getting serious attention. The FM frequency keeps your signal from getting lost in the multitudes of online choices. But most important of all, radio keeps the voices of your community saturating the airwaves, freely accessible for everyone within it to hear.

Why now?

Nonprofit organizations cannot compete in frequency bidding wars with national radio networks, and therefore they find little space on the FM dial. The FCC does not provide local nonprofit groups with opportunities to get on the airwaves very often. The rare opportunity to do so is the result of a decades-long battle by individuals and organizations who understand the importance of public access to the airwaves (outlined earlier in this chapter).

It's remarkable that communities now have a chance to build a radio station and get it on the terrestrial dial. However, there are not many spaces left on the FM band, especially in urban areas. If we don't claim and fill them with the voices of the people, they will be lost into the abyss of corporate repeaters. Chapter 3 tells you more about applying for one of those rare spaces.

Why you?

If not you, then who? First of all, not too many people know about this opportunity. With this book, upcoming conferences, and country-wide campaigns, this is something the grassroots radio community is trying to change. We need to get the word out and hopefully you will help.

The fact that you know about the LPFM movement places some responsibility upon you to take advantage of the situation. If you don't, and no one does, then all the activism and hard work will eventually have been for nothing. Radio will continue to create its own demise, and down with it will go our freedom of speech.

If you step up to this challenge, however, you will do so not only on behalf of your own community, but also that of people nationwide whose very freedom depends on open avenues for making their voices heard. In short, you should do this! The wonderful thing is, you don't have to do it alone. This book is your first step towards joining the radio revolution.

Creating Your FM Dreamscape

REMEMBER

Low Power FM radio is by its very nature a team effort. The licenses are awarded to organizations not to individuals and they should be developed as such. A healthy station finds the perfect balance between management and self-governance. Ruling with an iron fist will never work with LPFM. Chapter 12 provides advice on building a great team that can help your station grow.

The American public owns the American airwaves. LPFM frequencies are licensed by the federal government to nonprofits for a certain period of time. That time must be used wisely or future renewals will be denied. Although building a station should be a collective effort, it does make sense for those who put in the work of bringing frequency to life to maintain rules and expectations for those who get involved.

WARNING

Although it can take years to get an FM frequency, it can take just moments to lose it. Violating FCC rules can result in fines that are too substantial for most nonprofits to withstand. While sharing the airwaves with your community, you must be cautious about who you let on board, make sure that you train them well, and catch them immediately when mistakes are made. At the same time, you can't be afraid to take chances on unique personalities — or their big ideas. The FCC spends its time pursuing blatant violations of federal law. Don't freak about minor accidents, just make a correction and keep moving forward. Chapters 10 and 11 help you stay within the lines while pushing the boundaries.

Radio never sleeps. The most unbelievable things will happen. There is simply no way to fully prepare. You just have to be able to think on your feet and roll with all the punches. Chapter 13 provides some problem-solving advice.

Aside from that, the most important skill you can have is the ability to delegate. Of course, part of delegating is knowing who is up for the task. It is also knowing how to keep people motivated, even when there is no money involved. You must be nimble and flexible. Volunteers are reliable, until they are not. You must know every job and always be prepared to pick up the slack. Chapter 12 helps you master these tasks.

Community radio is fertile ground for new talent. It is a space where people with little media experience can get their hands on the equipment and try out new ideas. Chapter 11 provides creative tips for developing ideas into radio programs. Sometimes these ideas work, and sometimes they fail miserably. Programs that sound good on paper may not always live up to expectations, and shows that sound dull may light up the airwaves with surprise characteristics. You may find that people with fun personalities in real life have no vibrancy on the mic at all. And vice versa: Introverted individuals often light up on air and unleash a personality they keep locked inside. It goes to show that you never can tell.

TIP

Have a vision for your station and always keep it in sight. Perhaps you are an activist building a station to spotlight community issues and broadcast from local events. Maybe you are a musician who wants to highlight the music of your region and host on-air performances. Perhaps you represent a cultural group looking to broadcast health and voter information in multiple languages. Perhaps you are an athlete who wants to broadcast local sporting events. Maybe you are part of an artist collective that wants to stretch the FM possibilities to the outer limits. Perhaps you want to do all of this — and more — with a new free-form frequency. There are so many ways to envision a LPFM station. It is truly an open canvas, and the beautiful thing is that there is no right or wrong way to approach it.

Although there are some laws to follow, your own organization dictates the content on your station. Get creative! Chapter 16 suggests unusual ways to fill the air, and Chapter 17 will inspire you with some amazing moments in American radio broadcasting.

No matter how much you plan, your station will become different things at different times. You'll be astounded at how much it can change depending on what is happening in the world or which DJ is on the air. Holidays, for instance, are always fun to celebrate on the radio (Chapter 11 helps stir on-air programming ideas). Special shows will fill the airwaves, and suddenly your station may take on a whole new character. Emergency situations, of course, become a lot more intense (Chapter 4 looks at the Emergency Alert System you must install for such situations). As important as it is to have a vision for your station, it is also important to allow your station the flexibility to be all the things it wants and needs to be.

Raising a station is very much like raising a child. It's a lot of guesswork and a lot of improvising, but you will figure it out as you move along. Every organization will do it their own way and each young station will emerge with its own unique personality. I can't wait to hear about them all! You will have visions for your station and it will take on a mind of its own. Like a child, a radio station takes all the time, energy, and love you have, but no matter how exhausted you become, you still want to give it more. Chapters 8 and 9 help you understand what kind of resources your station will need to thrive while providing creative fundraising activities to help you acquire them.

You will love watching your station grow and will understand that the rewards are greater than the sacrifices. The amazing moments will make you forget all the struggles and you will know the true joy of seeing your dreams come true.

Challenges will be there for sure, but serve your listeners and they will serve you in exchange. Don't be afraid to call on them for the support, supplies, and donations that keep you going.

Make your station comfortable and welcoming and it will become a meeting place for your community. A place where friendships are made and collaborations are formed! Projects will take flight! Musical ensembles, theater troupes, and lifelong partnerships will be forged within your studio walls! Plans will be laid and ideas will be hatched! Magnificent things will happen! You will tune in, sit back, and wonder how it all came to be.

And then you will remember — it all started with you!

» Joining with an existing nonprofit or setting up your own

» Building your presence

Chapter 2

Finding or Founding Your Nonprofit

The Federal Communications Commission (FCC) only grants Low Power FM radio licenses to nonprofit organizations, public safety organizations, government agencies within their jurisdiction, and federally recognized tribal groups. LPFM stations cannot be individually held. Charitable organizations must be state registered nonprofits, but they do not have Federal 501(c)3 tax-exempt status.

If you are involved with one of these entities, that's a great start! The FCC has certain requirements for nonprofit organizations and their board members who wish to apply for a LPFM license, and prior to applying you will need to be aligned with an organization that qualifies.

If you are not presently involved with any of the above entities, you will need to find one or found one before applying for an LPFM license. In this chapter you'll discover what an organization needs to qualify and how to make sure that your group is in the best position to apply when the time arrives.

Preparing for the LPFM Application Window

To start broadcasting your LPFM radio station, you need to get permission to do so. And to get that permission, you need to apply for a license. Applications for a Low Power FM Construction Permit can only be submitted when the FCC opens an *application window*. The window is typically a one-week period during which entities may submit applications through the FCC website. The FCC promises that they will provide at least 30 days advance notice on their website prior to opening an LPFM window, although they often give a longer notice period. Even so, don't wait for the official announcement to start getting organized . . . if you are interested in applying for a LPFM permit, you need to get ready now!

Meeting LCRA expectations

The *Local Community Radio Act of 2010* (LCRA) created the rules for American Low Power stations (find out more about the Act in Chapter 1). Your LPFM application can and will be dismissed if your organization fails to meet any of the requirements set forth by the LCRA.

The full text of the LCRA can be found online at `https://www.congress.gov/111/plaws/publ371/PLAW-111publ371.pdf`. The FCC has distilled this information into an online guide to the Low Power FM service. This online manual includes information about the LPFM license process as well as answers to frequently asked questions. It is accessible at `www.fcc.gov/media/radio/lpfm`.

Keeping it local

According to the LCRA guidelines, only *local* groups can apply for LPFM frequencies. That means that your organizational headquarters needs to be within a 10-mile radius of your proposed transmission tower. If not, 75 percent of your board must live within that range. This radius increases to 20 miles for areas outside of the top-50 Nielsen radio markets (Nielson Audio is a firm that provides metrics on American radio markets). A list of the counties that are within the top 50 markets is available on the RecNetworks website: `recnet.com/metro`.

Upholding board expectations

Prior to filing your LPFM application, review the following information to be certain that nothing in the background of your organization will jeopardize your chance at gaining a license. Here are the things you need to be aware of:

» 80 percent or more of the board of directors must be American citizens.

» Board members and their families cannot have ownership interest in any other broadcast licenses, daily newspapers, or cable TV systems.

» Anyone who has been dishonest with the FCC in previous proceedings may not serve on the board of a LPFM.

» Directors may not have been convicted of felonies for crimes involving discrimination or providing fraudulent statements to the U.S. government.

» Directors may not have a conviction that has resulted in the loss of federal benefits under the U.S. Anti-Drug Act.

» Directors cannot have engaged in the operation of an unlicensed broadcast station.

Although a written statement from each of the board members attesting that they meet the above qualifications is not required by the FCC, it doesn't hurt to have a copy for your own files. You can find a board qualifications pledge in this book's online resources. Having your directors sign this document is an easy way to ask them about potentially uncomfortable situations. Directors who have conflicts or do not want to sign this document should resign from the board and look for other ways that they can serve the organization.

ASKING FOR HELP

The information you need to follow is pretty straightforward but some parts can be tricky to get your head around. To make sure that you are doing everything right, you may want to get some professional advice. Common Frequency (`commonfrequency.org`) offers free consultations and sliding scale guidance to organizations interested in applying for a LPFM license. The Prometheus Radio Project provides a wealth of LPFM resources including a helpful checklist for new radio stations on their website, `prometheusradio.org`. Both of these nonprofit organizations were involved with the passage of the *Local Community Radio Act* and know what it takes to help get your station on the air. REC Networks (`recnet.com`) is a private engineering company that has also been involved in the LPFM movement since the very beginning, continuing to provide a wealth of free information about the FCC rules and expectations on their website `recnet.com`. Brown Broadcast Services (`brownbroadcast.com`), Community Media Assistance Project (`c-map.org`), International Media Action (imarad.io), and Austin Airwaves (`austinairwaves.blogspot.com`) also have extensive experience getting new stations on air, even in the most challenging situations.

Scoring Priority Points

In urban areas, most LPFM frequencies will be competitive. It is likely that several organizations will be applying for the same spot on the dial. If they cannot agree to share the frequency, the FCC will use a point-based system to determine which applicants have priority. These points are something to keep in mind when finding or founding your nonprofit in urban areas.

TECHNICAL STUFF

One Priority Point will be awarded for each of the following:

- **Established community presence:** Two years or more of charitable work in the community.
- **Local program origination:** A pledge to create local programming for at least 8 hours a day.
- **Main studio:** A pledge to have the main studio. within 10 miles of the transmitter, and staffed and publicly accessible at least 20 hours a week between 7 a.m. and 10 p.m. To get this point, applicants must provide a phone number and address for the proposed main studio when filing their application.
- **Local program origination *and* main studio:** A bonus point for having both of the above.
- **Diversity of ownership:** Applicants host no attributable interest in other broadcast stations.
- **Tribal applicants serving tribal lands:** To qualify, 51 percent or more of the organization must be controlled by a tribe or tribes.

If there is no competition for a frequency, the point system will not need to be employed in acquiring the Construction Permit, so the LPFM licensee does not need to uphold the above expectations.

DIVERSITY OF OWNERSHIP — EXCEPTIONS TO THE RULE

For the most part, organizations can only apply for one LPFM frequency, with the possibility of up to two future FM *translators* (translators are separate licenses that rebroadcast the LPFM signal to extend the coverage area). A few exceptions exist:

- **Tribal applicants:** Tribal organizations on tribal land may hold two LPFM stations and up to four translators.

- **Public safety organizations:** In instances where there is no other competition, public safety organizations run by government agencies may apply for additional frequencies within their jurisdiction.
- **Accredited educational institutions:** Accredited educational institutions with an attributable interest in a non-student-run broadcast station can apply for a LPFM license that will be operated and managed by the students of the institution.

Uniting for Good: Partnering with an Existing Nonprofit

If your nonprofit organization has been incorporated for over two years, you qualify for a Priority Point (detailed in the previous section) automatically. If you represent a young nonprofit organization that has been incorporated for less than two years, you may be able to claim the Priority Point by proving that your group has been active in the community prior to its time of incorporation. Newspaper articles and event flyers with the date and your organization's name on it are items that can be submitted to the FCC as evidence.

You may have to develop some strategy in order to claim your community presence Priority Point. For example, consider partnering with an older nonprofit until yours comes of age. This organization can apply for the LPFM frequency when the window opens and you can manage it for them until you reach the magical two-year mark. At that point, your partner can transfer the license over to you using FCC Form 314 (discover more about this process in the Chapter 3).

REMEMBER

Partnering with an existing nonprofit comes with its rewards and pitfalls. The obvious, and very big, plus is the two-year Priority Point. It can also save you a lot of bureaucratic work of incorporating, filing taxes, and so on. Your partner organization may be able to provide studio or antenna space, administrative assistance, and volunteer DJs. If you choose this option, however, it is imperative that your vision for the station aligns with that of the partnering nonprofit. Their work, their donors, their decisions will all reflect on your organization and vice versa.

WARNING

In such a partnership it is important to have a written document that outlines who will be in charge of station finances, operational maintenance, and creative decisions. If the group will impose any limitations on the content of the broadcast, it will be important to know that before getting involved. Since the board of directors of the organization that holds the license can decide to transfer it or turn it back into the FCC at any point, it is imperative to be extremely careful if you go the partnership route.

Starting from Scratch: Forming a New Nonprofit

Nonprofit incorporation can be achieved quickly and without great expense. The speed with which incorporation can be accomplished compared with the slow pace at which the FCC moves can be helpful in earning the two-year Priority Point discussed earlier in this chapter. Since you have control over when you incorporate, but have no control over when the FCC opens the window, incorporate as soon as your organization is able to meet your state's requirements.

There's a lot to learn when you set up your own nonprofit, but you can do it by taking it one step at a time. The following sections discuss the key steps to take.

Appointing your board of directors

REMEMBER

Anyone can start a nonprofit (not-for-profit) organization. It all begins with a board of directors. A nonprofit *board of directors* (comprising a team of board members) is responsible for establishing and maintaining the Mission of the organization (see later in this chapter for more on your Mission). They make sure that the organization has the financial resources available to fulfill its goals. Board members are not owners of the organization; the law forbids board members from having any ownership in the profits of the nonprofit organization.

The nonprofit board works on behalf of the public to make sure that your organization is living up to its promises to the community. The board develops organizational policy. They hire a general manager (see more about staff positions in Chapter 12) and subsequently support and supervise the manager's work, which includes the hiring and firing of other members of the staff. Each board member serves as an ambassador representing the organization and cultivating relationships that will help it thrive. Board members develop strategic plans for the organization that include self-evaluation and succession planning.

The fundamental legal duties of individual board members include

- **Duty of care:** Actively participating in organizational decision making and using good judgment when so doing.
- **Duty of loyalty:** Pledging that decisions are made with the organization's best interest in mind; personal interests are set aside.
- **Duty of obedience:** Ensuring that the organization abides by all applicable laws and regulations, including its own by-laws (see later in this chapter for more on these).

In most states, you will need a minimum of three individuals to get your organization off the ground. A few states only require one. Founding board members can be friends or community members, but make sure that they are people you trust. They should believe in your Mission and have an interest in cultivating community radio. They should be easy to work with and feel motivated to help. Get to know potential candidates through meetings or volunteer work before offering them a spot on the board, and don't bring anyone onboard whose vision does not align with the goals of the organization. Ideally, the board meetings will be an enjoyable exchange of ideas among a talented group of problem solvers. You need to be able to trust and confide in every member.

Don't take on too many board members. Seven is likely enough, though many boards have more. Having multiple voices in the room is good but not to the extent that meetings become difficult to manage. Keeping an odd number makes voting easier in case not every decision is unanimous.

Make sure there are no conflicts of interests among your board members. The IRS requires a *conflict of interest policy* that must state that board members disclose when they have (or think they have) a personal interest that may affect the decisions of the organization, and prohibits board members from voting on any matter that may affect their personal interests. Many organizations have their board members sign an agreement to uphold the conflict of interest policy.

Potential board members will want to know the following: How long are the terms? Are there financial obligations? How often will the board meet? Will they be expected to fundraise, plan events, serve on committees, and so on? It's okay if you don't know these answers right away; your founding board members will make many of these decisions as they go along.

Think about diversity on your board and make sure that the directors of your LPFM station reflect the demographics of your community. Women and people of color have traditionally been excluded from radio leadership roles as well as nonprofit board positions. Now is your opportunity to make up the difference!

In the beginning you may have a working board that is more involved with the daily activities of the station. Think hard about the skills you need to run your organization and try to populate the organization accordingly. Attorneys, accountants, graphic designers, event planners, sound engineers, and IT professionals can all be very useful to a new organization. Radio experience can be helpful but is not required by the FCC.

It's a good idea to have a plan or at least an outline of a plan to transition the board from a *working board* to an *oversight board*. This may not be in the form of a written document (though that's not a bad idea), but your young board should keep this in mind as the organization matures. During this process, the board of directors should shift its focus to big-picture goals for the station while allowing the station management to handle day-to-day operations. That said, the board should provide support whenever it is needed. They should be a springboard for new ideas and an advice network for handling challenging issues.

TIP

Board members should be celebrated as often as possible for their role in founding the station and leading its course of development. Their names should be included on the station website and in printed materials. Providing food and beverages at each meeting can make everyone feel ready to spend some time doing the work. Invite board members to the station's social events and introduce them to all the DJs. It is wonderful listening to them share ideas! The board is a critical part of your team, so it is super important that they feel appreciated.

Hosting the initial board meeting

After you have selected your board of directors, you need to call a meeting in which you define the organization as well as its goals and aspirations. At this meeting you must

- Adopt by-laws
- Elect any new board members
- Elect officials

As your organization grows, establish board committees dealing with particular areas of station development. Set board meeting dates as soon as possible to make room on everyone's calendar. The board should meet a minimum of once a year, in-person or virtually. At ART FM we find that meeting every two months, rotating between in-person and virtual, makes for a good schedule.

Selecting board leadership

All nonprofits must have a president, treasurer, and secretary. It varies by state whether the same individual can hold more than one of these positions.

Every organization needs leaders who are making sure that things are running smoothly and getting done correctly. The following roles are standard for nonprofit boards:

- **President (or chair):** Leads the board of directors in making sure the organization is upholding its mission, supporting its staff, and maintaining its fiscal and legal responsibilities.

- **Vice president (or vice chair):** Accepts tasks and responsibilities from the president. Steps into the top leadership position when the president is not available.
- **Secretary:** Maintains organizational records, records meeting minutes, documents votes, and helps the board uphold organizational by-laws.
- **Treasurer:** Responsible for fiscal oversight of the organization. Prepares, upholds, and distributes budgets and financial reports.

It is possible for the station founder or general manager to serve on the board of directors and even serve as the president. It is reasonable that they would want to steer the organization that they have established. They should, however, be certain to recuse themselves from decisions regarding their own performance, pay rate, and similar details.

Naming your organization

You need to state the name of your organization in your Articles of Incorporation (defined later in the section, "Incorporating your nonprofit"). Perhaps you have a name for your organization already or maybe you need to make one up. If not, don't get stuck and don't waste valuable time. If you are going for the two-year bonus point mentioned earlier in this chapter, you need to get established as soon as possible. Throw a dart at the dictionary if you have to; you can always change the name of your organization through some simple paperwork later on through your secretary of state.

WARNING

It is illegal to name your nonprofit organization in such a way that it could be misleading. Words such as "Federal" and "United States" are not allowed. You also may not use words such as "Bank," "Insurance," or "Engineer," if you are not qualified to hold those titles. Some states require a designator after the organizational name such as "Inc." These rules vary by state; do some investigating with your secretary of state before submitting your Articles of Incorporation.

On your secretary of state's website you will be able to search their listings to be sure that there are no other companies using your intended name. Before locking things down, search online to see what comes up. Check to see if you can get the web address you want or if there are any other preexisting pages that could make finding your potential home page difficult.

Defining your Mission

After the board has been identified, they will work together to draft a Mission Statement that should be specific enough to keep your organization on task yet broad enough to allow the operations to grow and develop in unexpected ways. You will eventually need to prove to the FCC that you have an educational purpose for applying for an FM frequency. This does not have to be in your Mission Statement but it certainly would not hurt to have it there.

REMEMBER

Your Mission Statement will be the guiding light of your organization, so it is important to get it right. It is usually a concise statement that defines the organization's culture, values, and reason for being. The purpose of the Mission Statement is to keep the board, staff, and volunteers working toward a common goal. It helps those outside of the organization understand what it does and who it intends to serve.

Here are three examples:

- **ART FM / WXOX Louisville:** *Cultivating educational opportunities and live media experiences, ART FM employs sound, music, and conversation to amplify the local arts community and broadcast the international arts dialogue.*
- **Friends of WPVM, Inc:** *Our mission is to create radio programming that cultivates dialogue, informs, entertains and raises local, regional and global community awareness.*
- **The Voice of Arkansas Minority Advocacy Council / KLEK Jonesboro:** *Community empowerment through consistent outreach and engagement.*

Focusing your vision

Creating a Vision Statement is your opportunity to dream of a perfect future where all your organizational dreams come true! This is your time to put on your rose-colored glasses and think about where your organization would be if it lived up to its Mission and fulfilled all of its goals. This Vision Statement should give your volunteers and employees something to aim for. It should be inspiring and uplifting. It paints a picture of an exciting future for the organization and all those who benefit from its services.

For example, KMRD-LP in Madrid, New Mexico, has the following Vision Statement:

> ***KMRD*** *serves the local public by being an accessible and inclusive resource for entertainment, knowledge sharing, community organizing, and public safety.*
>
> ***KMRD*** *builds community by amplifying diverse voices, facilitating constructive communication, and fostering meaningful connections.*
>
> ***KMRD*** *uses participatory radio to promote positive change by improving local self-sufficiency, enhancing community well-being and resilience, and promoting social inclusion and economic justice.*

Incorporating your nonprofit

Articles of Incorporation are the documents used to establish your organization as a legal entity. Drafting your Articles of Incorporation and submitting them to the proper authorities can be done quickly and easily with the help of a legal professional. It is a straightforward process and most law offices will offer a flat fee for this kind of work.

TIP

If you do not know an attorney, speak to legal aid societies, nonprofit support organizations, and arts agencies in your area. They may be able to provide helpful contacts or legal assistance. The IRS provides some suggested language for incorporation on their website: www.irs.gov/charities-non-profits/suggested-language-for-corporations-and-associations.

The following items will be required in your Articles of Incorporation. A few suggested answers are included, but your specific details may be different (consult your attorney):

- Legal name of the organization
- Address of the organization
- Name and address of registered agent (the representative to receive legal processes for the organization)
- Duration of organization (perpetual)
- Statement of purpose
- Board of directors (full names and addresses)
- Membership (or no members)
- Amendments (if applicable)
- Name and signature of incorporator (the person filing the documents)

WARNING

You need to ensure that all of your organization's work serves its stated charitable purpose. To that end, the IRS strictly prohibits campaign intervention by Federal 501(c)3 tax-exempt organizations; you can read the full details here: www.irs.gov/charities-non-profits/charitable-organizations/the-restriction-of-political-campaign-intervention-by-section-501c3-tax-exempt-organizations.

The IRS also requires a *dissolution plan* in which the assets of the organization will go to another nonprofit or to a government agency serving the public good should the organization decide to cease operations.

After the articles are drafted, each of your initial board members will have to sign the document. You or your attorney then files these with the secretary of state. As soon as your nonprofit filing is approved, which should take just a few days, you will be qualified to apply for a LPFM license (discussed in Chapter 3).

Drafting your by-laws

Organizational *by-laws* provide specific instructions for how your organization should act internally. The organization is legally required to follow these by-laws, so keep the requirements they impose and the processes they establish as simple and streamlined as possible. Write the imperative things in there but leave out the rest, and review the nonprofit laws within your state, with which you also will be expected to comply. An attorney will be very helpful in drafting by-laws that keep your organization focused on its Mission but flexible for future opportunities.

Items that should be in your by-laws include

- Number of board members (how they are elected and dismissed)
- Directors and officers (roles and duties)
- Definition of a *quorum* (the minimum number of members that must be in attendance to make a meeting and its decisions valid)
- Rules and procedures for meetings, votes, and elections
- How investments and properties will be managed
- Nondiscrimination policy
- Conflict of interest policy
- Methods for amending by-laws

TECHNICAL STUFF

Nonprofit by-laws are not public documents, but the board should review them often to be sure that they are working in compliance and, if necessary, amend them accordingly. Keep in mind that the IRS requires Federal 501(c)3 tax-exempt organizations to report any "operational changes" in their subsequent 990 tax filing. This includes updates to your by-laws (see later in this chapter for more on taxes).

Applying for your Employer Identification Number

After your organization is legally formed, you will need to get a *Federal Employer Identification Number* (EIN). This is the number you will use to conduct official business such as annual tax filings and to apply for your tax-exempt status. EIN applications and instructions are available on the IRS website (irs.gov). Select "Apply for an Employer ID Number" and then click through for the "Other Non-Profit/Tax-Exempt" option.

Establishing and maintaining 501(c)3 status

The term *501(c)3* is used widely to refer to tax-exempt organizations. The term refers to section 501(c)3 of the U.S. tax code that defines tax-exempt organizations and their Federal obligations (namely that the organization work for charitable purposes, and none of its earnings will benefit any individual shareholder). The 501(c)3 code also states that the organization may not be an "action organization" that influences any legislation or campaigns for or against any political party.

It is not necessary to get 501(c)3 status to hold a LPFM license. However, doing so will be a financial advantage to your organization. It allows donors to deduct gifts made to your organization from their annual taxes. Accordingly, there are a number of businesses, individuals, and granting agencies that only donate to 501(c)3 organizations.

To become a federally recognized tax-exempt organization you will first need to file a Form 1023 "Application for Recognition of Exemption under Section 501(c)3 of the Internal Revenue Code" with the IRS. After they are submitted you will need to wait for approval, which could take up to six months. There is an expedited application form and review process. A 1023-EZ is available for organizations with less than $250,000 in assets and an expected annual budget of less than $50,000 for the next three years. For all others, the 1023 must be used and it is longer, requiring more financial and operational information than the EZ. You can find out more at www.irs.gov.

If you do not qualify for the EZ, you will need to submit copies of your Articles of Incorporation, by-laws, and conflict of interest policy to the IRS with your 1023 application. To avoid complications and delays in processing your 501(c)3 organization, I recommend hiring an attorney to assist. The paperwork can be between 50 and 100 pages long but should be a fairly straightforward process for a legal professional. They can usually handle the entire process for a reasonable rate on top of the mandatory $600 Federal filing fee.

After the IRS approves your application, you will receive an official letter recognizing your 501(c)3 status. Make several copies of this document as you will need it for additional filings such as your state sales tax exemption (discussed in the next section). It can take up to six months after filing your 501(c)3 application with the IRS before your organization is granted tax-exempt status. However, gifts made to your organization while the application is pending will be retroactively tax-deductible when you finally receive your tax exemption.

TIP

The IRS offers online workshops that can help answer questions and navigate the application process. You can learn about the IRS and their online educational resources at `www.irs.gov/charities-non-profits/application-for-recognition-of-exemption`.

Getting a state sales tax exemption

As soon as your organization becomes a Federal 501(c)3 organization, you can send a copy of this authorization to your secretary of state for an official exemption from state sales tax. After processing, they will send you a state sales tax-exemption number. Make a few copies of this letter and be sure to keep one on file. If you present this letter to businesses that operate within the state, you will not be required to pay sales tax on any organizational purchases. You can use this number to register for tax-exempt accounts with online retailers, including Amazon, eBay, and Reverb.

Filing 990s

When you start a nonprofit organization, you will be responsible for regular business filings including federal taxes. Nonprofits with less than $50,000 in gross receipts can file a 990-N (e-Postcard). This is an extremely simple form that takes less than 10 minutes to complete. Organizations with gross receipts between $50,000 and $200,000 can file the 990-EZ. This is a little more complex but you may still be able to complete this abbreviated tax filing yourself. Having a professional review your work, however, is recommended and this is where it is helpful to have a CPA on the board of directors! After your organization is making $200,000 or more a year, you will have to complete the standard 990. This paperwork gets much more complicated and I recommend using some of your six-figure budget to hire an accountant to handle your tax reporting.

WARNING

Nonprofit tax filings are due by the 15th day of the fifth month after the organization's accounting period ends. This is May 15th if you work on a calendar year. Failure to file your nonprofit tax return by the scheduled deadline can result in fines of $20 per day which can add up very quickly. Organizations that fail to file their nonprofit tax returns for three years will have their 501(c)3 status automatically revoked. Make sure that you know the deadlines and get your paperwork filed on time!

Each city and state has different annual filing demands for nonprofit organizations. Most states require nonprofits to file an annual report with their secretary of state, which can be as simple as a declaration of board members. Check with your state and local business agencies to be sure that you are completing all the expected paperwork.

Obtaining nonprofit postage

The U.S. Postal Service offers reduced postage for federal tax-exempt organizations. To obtain these reduced rates you must apply for a nonprofit permit at your local USPS Business Mail Entry Unit. You will need to complete the PS Form 3624 "Application to Mail at Nonprofit USPS Marketing Mail Prices." This form requires that you present a copy of your Articles of Incorporation, your 501(c)3 determination letter, and other evidence of nonprofit status such as an official financial statement. Additional materials such as newspaper articles, organizational brochures, and member bulletins that attest to the organization's charitable activities within the past 12 months may be submitted as supportive materials. There is no fee to apply for nonprofit postage authorization. It should take about two weeks to get approval.

After it is approved you activate your nonprofit mailing license by paying an annual fee of around $300. Your license is active for one year from the time of payment. You may let this lapse and pay again when you are ready. The one-year clock will start over at that time. Beware, however, that if you fail to make a nonprofit mailing at least once every two years, you may lose your authorization.

All told, it can be an investment to save money on your nonprofit mailings. Before getting started, you will have to think about how much mail you will be sending and calculate whether it is worth it for your organization. The reduced postage rates are less than half the cost of a regular stamp but they only work on bulk mailings of 200 or more on letters or postcards prepared according USPS details. There is no discount on individual mail pieces or packages.

TIP

Instructions for preparing nonprofit mail is available on the USPS website: `https://pe.ups.com`.

Protecting yourself and your assets

One of your largest expenses when starting a nonprofit is your insurance payments. As soon as you begin your operations you will need business liability insurance. After you acquire any assets for your business you will need to protect those as well. Property and liability policies are usually bundled together.

Many nonprofit organizations also take out Directors and Officers (D&O) insurance. This protects individual board members against lawsuits that may result on account of their involvement with the organization. Although D&O insurance may be prohibitively expensive for new organizations, individual board members may be able to add a similar protection to their homeowner's policy for just a few extra dollars a year. Board members should discuss this with their insurance agents and make the decisions that make them feel the most comfortable.

When your organization begins broadcasting you may want to also take out a *professional broadcaster's policy*. This will protect the organization from things said on air by the staff, volunteers, or any of their guests. This is extremely important for a community media organization where you just never know what is going to happen!

When shopping around be sure to let the agents know that you are a nonprofit organization. They may be able to waive certain taxes and fees.

Making connections and receiving funds

You will need avenues of contact for your organization such as an email address, a phone number, and a mailing address. Google Voice (`voice.google.com`) offers free online voicemail accounts that can serve as the organization's phone number as you get started. Getting a P.O. Box is a great option for a new nonprofit that does not yet have its own physical space. P.O. Boxes are usually quite affordable, but they cannot accept FedEx or UPS deliveries.

You'll also need a bank account. Shop around to find a bank that offers free checking accounts for nonprofits. It's a good strategy to work with a local bank known for supporting community activities. Then, set up PayPal, Venmo, or similar payment accounts as soon as you can — make it easy for supporters to give money right away!

Try to get all these methods of communication and payment up and running as soon as possible.

Making Your Mark: Branding Your Organization

First impressions are everything — you need to use your public image to attract listeners to your broadcast. Your board of directors should brainstorm many ideas before settling on the look and feel of your organization. This section looks at the basics to get you started.

Developing a logo

You want your logo to stand out. It should be striking enough that it attracts attention and attractive enough that folks will want show it off. Clothes, cars, and drum kits — you want to see your sticker everywhere! You can design your own logo or hire a local artist to bring it to life for you. Get started by collecting ideas and graphics that appeal to you; think about a palate of station colors. Figure 2-1 shows a variety of logos from LPFM parent organizations. You can see the variety of stylistic approaches.

REMEMBER

Do your due diligence. Search online for similar names and logos used by other entities in radio and communications. Adopting a name or logo that is too similar to one already in use can be a trademark violation. Consult with an attorney before you start using and promoting your station and investing in making your name and logo known; don't fall in love with a name or logo until you know you can own it. With original artwork, make sure to obtain a copyright agreement in writing from the artist. It will be important that you have unlimited use of the logo and the right to modify it as necessary.

FIGURE 2-1: Logos from several LPFM parent organizations.

Building a website

No matter how simple it is, you will need a website. At the very least, include your logo, contact information, and a donate link. After you get broadcasting it should feature your call letters and frequency prominently. If you stream online (see Chapter 5), the player button should be obvious and easy to use, and be sure to include links to social media pages where fans can get connected with your cause. These elements are demonstrated on the ART FM website shown in Figure 2-2. Without a doubt, your web address should be included on every single piece of station merch and printed material you produce!

TECHNICAL STUFF

If you can build your own website using HTML code, please do so! I am always in favor of open-source publishing. For those who are not fluent in programming language, templates such as WordPress and Squarespace are great alternatives. Principal managers of the station should be familiar with web-design processes and should be able to actively access and update the pages. Relying too much on others to design your site can be disastrous should the volunteer quit or the contractor become too expensive.

Establishing a presence on social media

Social media is a wild, constantly evolving landscape, and whatever form it takes, now and in the future, your organization will need to be there sooner rather than later: Facebook, X (formerly Twitter), Instagram, TikTok, YouTube, Discord, and more. Radio is often a one-way street but social media allows you to hear back from your listeners. Social media is a fantastic way to let the world know about upcoming events, special programs, and station fundraisers. Get creative on social media and find yourself connecting with a world of international listeners!

Building community connections

Get out there! Introduce yourself to your neighbors. Host open houses, involve yourself with organizations that support nonprofits, attend courses and social activities. Connect with other local nonprofits and show up at community events, planting seeds as you go, cultivating new radio shows and topics of on-air conversation. Whether you host a table, provide a radio demonstration, or just show up with a fistful of stickers will depend on the amount of time you can give. Just get out there however you can! Now that you have built an organization, you've got to let people know it is there, what it is all about, and how they can get involved!

FIGURE 2-2: The ART FM website has the following important elements: player button, donate button, and links to social media pages. The page includes recent information, donor acknowledgments, links to the schedule, program pages, station history, listener surveys, and upcoming events.

- » **Locating a tower site to broadcast from**
- » **Completing your LPFM application**
- » **Investigating other FM opportunities**

Chapter 3
Getting Permission to Broadcast

This chapter tackles your first steps towards getting on air after you've set up your nonprofit organization as covered in the previous chapter.

During the Low Power FM application window — more on this important event later in the chapter — nonprofit organizations have the opportunity to apply for empty spaces on the FM band. To get broadcasting, you need to find your spot on the dial! Prior to granting approval for a new station, the Federal Communications Commission (FCC) wants assurance that the additional broadcast will not create interference with stations that are already on the air.

In this chapter you find out how to search for channels in your area, find a broadcast location, file your FCC paperwork, defeat your competition, and claim your spot on the FM dial!

Finding Your Frequency

FM (Frequency Modulation) frequencies, the beachfront property of the radio dial, can be extremely hard to come by and very expensive when you do. Although airspace may be readily available in rural areas, it can be nearly impossible to find in large urban zones (unless you have very deep pockets).

Thankfully, the passing of the *Low Power Community Radio Act of 2010* (LCRA) made it possible for local nonprofits to apply for small spots on the FM dial, to squeeze in between larger broadcast channels. Unfortunately, these small frequencies may not be available in all cities. It all depends on the density of the FM band. It is, however, worth a shot to investigate in your area. Frequencies do open up in unexpected places from time to time and there may be other creative solutions.

In most cases, nonprofit organizations will only be allowed to apply for one LPFM channel. There are, however, special allowances in which universities, tribes, and government organizations can apply for additional frequencies if specific criteria are met and there's no other competition. See Chapter 2 for an understanding of the available opportunities and necessary qualifications. Visit the FCC website to get the official information on the LPFM service and the official updates on applying for a LPFM frequency: `www.fcc.gov/media/radio/lpfm`.

Since LPFM frequencies are specifically reserved for local organizations, your search will be limited to the area where your organization is based. Specifically, this means that the transmitter must be within 10 miles of the organizational headquarters or within 75 percent of the board members' homes. Outside of the Nielson Audio top 50 radio markets, this radius increases to 20 miles.

If possible, apply for a frequency that no one else is after. After you decide to apply for a frequency, discreetly try to determine if any other groups are applying for a LPFM license in your area. If so, find out what frequency they are going for. This may require a meeting between your two groups, and even then they may not want to tell you. Depending on where they are applying and what your options are, you may want to go for a different frequency than the one you initially had planned.

If there is a lot of competition in your area, it may benefit you to partner with a like-minded nonprofit that can share the frequency and beat out others with your combination of points (for more on totaling up points, see Chapter 2). If your combined group wins the license, they may be happy for you to manage it so long as they get a bit of airtime. Down the line, if they are not interested in running the station, they can hand their portion of the license over to your nonprofit organization using FCC Form 314.

Locating quiet spots on the dial

The FM band in the United States is divided into 100 channels beginning with 88.1 MHz and extending to 107.9 MHz. Scrolling through the band, you will notice that all the broadcast stations end with an odd decimal. They go 88.1 to 88.3 to 88.5, and so on. You can learn the about math behind the frequencies on the FCC website: `www.fcc.gov/media/radio/fm-frequencies-end-odd-decimal`.

To apply for a LPFM license, you must find a frequency in your area not currently being used by another broadcaster. This is known as your *co-channel*. Then you must make sure no other stations are broadcasting on the first adjacent channels; the *first adjacent channels* are the neighboring frequencies. If there are pre-existing stations on the first adjacents then the co-channel is not available. If there are no stations on the first adjacent channels, you have to take it one step further and see if anyone is broadcasting on the *second adjacent channels*. If there are pre-existing stations on one or both of the second adjacent channels, things get more complicated. In such a case, an engineering study (detailed in the next section) is required to prove that the new broadcast will not cause any interference with these pre-existing stations.

If this sounds a little confusing, don't worry because Figure 3-1 illustrates the channels that are adjacent to the co-channel 97.1 FM in an easy-to-follow way.

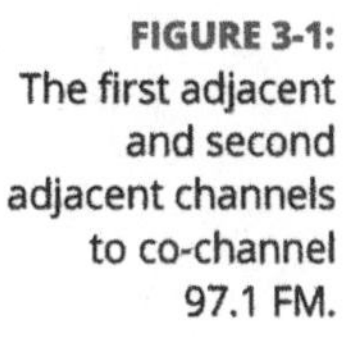

FIGURE 3-1: The first adjacent and second adjacent channels to co-channel 97.1 FM.

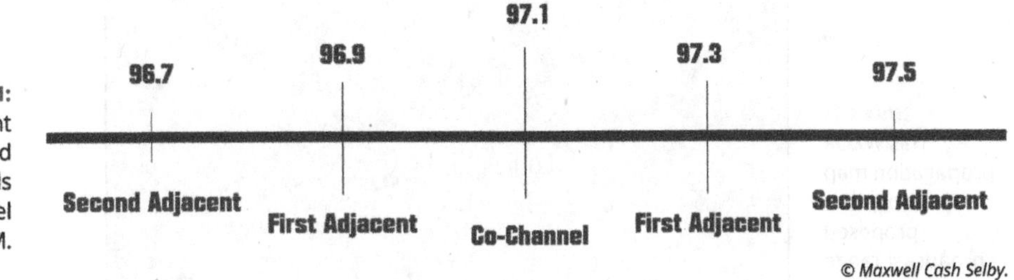

© Maxwell Cash Selby.

If you are lucky enough to find an open frequency in your area with no one broadcasting on the first adjacent or second adjacent channels, you are in good shape! You do not need to submit an engineering study with your LPFM application but it is a good idea to consult with broadcast professional to be sure.

Securing an engineering study

I recommend finding a broadcast consulting engineer who can help with your application and determine factors not under the FCC's jurisdiction, such as environmental and land use limitations, property availability, zoning, and FAA regulations.

Broadcast engineers at Common Frequency (`commonfrequency.org`), the Prometheus Radio Project (`prometheusradio.org`), International Media Action (`imarad.io`), REC Networks (`recnet.com`), and Brown Broadcast Services (`brownbroadcast.com`) will all conduct a free initial survey for your organization and work hard to help you find a frequency within your permissible area.

TECHNICAL STUFF

If you find a frequency with a second adjacent neighbor and you need an engineering study to submit with your application, your broadcast engineer will provide it for you for a fee. The engineer will review the contours of the landscape and the placement of the proposed LPFM antenna to predict the area where the new signal will reach. You can see an example of the WXOX contour map as provided by REC Networks in Figure 3-2. This information is compared to the contours of the second-adjacent broadcasters. The study needs to prove that the new LPFM broadcast will not create any new interference. Sometimes the engineer may recommend a directional antenna to prevent any contour overlap. If so, this information will be included in the study.

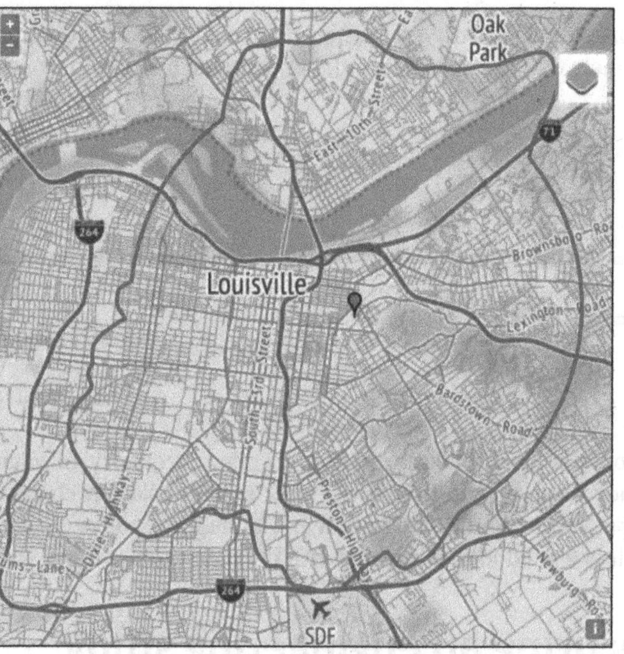

FIGURE 3-2: This WXOX propagation map shows the proposed broadcast range of WXOX-LP Louisville. The central location of the tower site provides coverage of the entire downtown area. Map tiles by Stamen Design, under CC BY 3.0.

Hunting for Your Tower

After you determine that there is an available frequency in your area, you should begin looking for a site to host your FM transmission. This means that you need to find somewhere to place an antenna — ideally on a tower. A *tower* is a mast that elevates your broadcast antenna away from regularly inhabited space and allows the broadcast signal to cover a greater distance.

If you are lucky, you will be able to add a tower to your studio location and that will make everything very easy. The LPFM antenna is neither very heavy nor must it be hoisted very high. In fact, it is much smaller than the kind of old TV antenna you still see on the top of some people's houses. Adding a mast to your own property may not be as difficult or as expensive as you think. Tower pieces can be found and assembled with the help of your local *ham radio* club (amateur radio enthusiasts). Pieces of aluminum fence railing also have been known to work.

WARNING

In order to prevent any radiation dangers, the radio antenna must be hung at least 20 feet above any occupied space. Failure to make this assurance will result in your application being disqualified.

You may need to find a remote location away from your on-air studio for your transmission equipment. It requires more legwork — and equipment! — but many radio stations need to do this. You can use point-to-point FM communications or a dedicated Internet service to get the audio from your studio over to a distant tower site; your broadcast engineer will provide you with the best options for your situation (discover more in Chapter 4 and in the online resources for this book).

Taking the first steps

The first thing to do when seeking a transmission site is to start driving around in your target area, looking for tall buildings or properties where you may be able to construct a tower. Be mindful that sites close to airports may not qualify due to possible interference with air traffic control.

TIP

Write down addresses of potential sites so you can then run them past your engineer to see which of these specific sites would work for your broadcast purposes.

Your tower site equipment must be secure and inaccessible to the general public. Yet you must have the ability to access the site when there are technical problems and any time during regular business hours that the FCC shows up to inspect.

You need to confirm that any proposed site has electricity, ideally access to an Internet service, and a place to house your equipment including an FM transmitter, remote computer, and EAS box (for more on this, see Chapter 4). If there's no housing, can you build a simple box on the site? If the location is off the grid, are you able to connect electricity services in some way? How about installing some solar panels? Depending on your location, you may or may not need climate control to keep your equipment from melting or freezing up. Your transmission equipment will produce some heat, which helps to keep it warm in the winter, but you will want to be sure everything is properly ventilated.

Internet service makes things much easier but it is possible to broadcast without it — as was done for many years — using point-to-point FM transmission to get your studio audio up to the antenna. Internet service may be worth its weight in gold, however. If you use it to access your transmitter and connected equipment remotely it can save you many late-night trips up to the tower. More on constructing your transmission site can be found in Chapter 4 and in this book's online resources.

FINDING A GREAT TOWER SITE

Todd Urick, Technical Director of Common Frequency, has the following words of wisdom to help you find an ideal tower site.

"Typically your choices are co-locating at an established tower site, co-locating on top of someone's else's structure, or building your own tower. For best coverage, in most cases, a height of 100 feet is sought. FM works best if the antenna is above the tree/building level.

In the first case, tower stock is divided into large tower firms and independent owners. In the last two decades, corporations like American Tower, SBA, and Crown Castle have procured many of the nation's towers — they all have convenient online site locator maps. It is important to realize these companies ask premium rates for co-location starting that could start at $500/month, and there are hidden one-time costs such as application processing, intermodulation study (to check if your antenna interferes with the antennas already there), and engineering loading study (to check structural capacity).

For a low-overhead LPFM, one traditionally seeks space for "free" to $250/month. With inflation, that has crept upward. You want to first see if there is an independent owner you can approach. Free-standing structures are primarily divided into *lattice towers* (free-standing), *guyed towers* anchored by "guy wires" to the ground, and *monopoles* (solid poles usually used by cell carriers). All are appropriate for FM antennas, but you are typically looking for space greater or equal to 10 vertical feet from another co-located antenna if available. "Stealth" cell monopoles (that look like other things, such as trees) are usually not co-locatable.

The website `www.antennasearch.com` provides a starting point for free-standing communications structures. A more limited selection of FCC registered towers can be seen by downloading the `FCCinfo.com` app for Google Earth."

Negotiating with a landlord

After you find a suitable site for your tower, the next step is to find the property owner and offer them the once-in-a-lifetime opportunity of having a radio station broadcasting live from their property. Very exciting! Any business would be foolish to pass up this grand opportunity for good public relations. By allowing an FM transmitter on their property, the business can reap the benefits of city-wide name recognition and a philanthropic reputation.

In exchange for their support, your programmers can thank the property's owner on air regularly and you can even provide them with tower naming rights. DJs can announce, "You are listening to WXOX radiating from the top of Occam's Tower in the heart of downtown Louisville" (obviously change the names to reflect your station and host). So long as this is done within FCC noncommercial guidelines (see Chapter 10 for more on this), and there are no endorsements or calls to action, DJs may thank the tower sponsor at regular intervals throughout the broadcast day. It won't take long for the sponsor to develop name recognition within the city and a positive reputation within the listening community.

After you find a landlord that is interested in hosting your FM antenna you should arrange a time to meet them, walk through the tower site, and share your ideas. As you speak with them, make sure they will be agreeable to the programming you plan to host. You don't need to lay all your cards on the table, but you should be able to get a good idea of this person's disposition — and their willingness to support your project — through a short conversation.

Look out for any controversies involving the tower's landlord and consider how these may reflect on your community broadcast. There are some institutions that have tried to restrict the programming on LPFM stations when they do not agree with their organizational morals. The Catholic Church, for instance, may have regulations against discussing certain sex-ed information. An ammunition company might have objections to a live debate on gun control. Although they may not have any legal means for governing your programming, they can also make things very uncomfortable if they hold the keys to your broadcast home. As always in life, it is important to be careful when entering partnerships — always make sure your values align.

Although you don't want to upset your landlord with your programming choices, you also can't let them disrupt your scheduled broadcasts. A contract will help keep things professional and mediate any trouble. You should find a property lawyer that can help draw up a simple rental contract. Even if the owner wants to donate the space, you should agree to rent it for one dollar a year. This kind of formal agreement ensures that you won't get kicked out of the space right after you get everything set up.

Sometimes you will find that your landlord has additional space they are willing to provide for a studio as well. Having the activity of a live radio studio on their property can be a real attraction, especially to a hotel, university, shopping complex, or entertainment venue.

Property owners need to be amenable to allowing the FCC access to the tower for inspection at any point during regular business hours. They also need to provide your technicians with reasonable access to the tower site as necessary to make technical repairs and updates. Thankfully, this can be kept to a minimum with the use of remote-control features in modern computers and transmitters (there's more about this in Chapter 4). After everything is installed correctly, you may go months without having to physically access the tower.

Applying for an FM Frequency

As I discuss in Chapter 2, you have to wait for the FCC to open an application window before applying for a LPFM Construction Permit. If there are no LPFM application windows on the horizon, you still have a few possibilities for getting on the FM dial but you will have to be creative. Find these ideas at the tail end of this chapter. Of course, you can begin broadcasting online at any time. It's a great way to get some practice and begin building an audience prior to getting on the dial. There is no authorization required to broadcast online and no FCC rules to follow.

The FCC announces application window dates on its website (`www.fcc.gov/media/radio/lpfm`) no less than 30 days in advance of its opening. The contact for questions concerning LPFM is the FCC Audio Division, found online at `fcc.gov/media/radio/audio-division`, or (202) 418-2700.

The window is open for only a brief time and you will not be able to apply before or after that window. Make sure that all of your information is in order and ready to file when the time arrives. Double-check that all documents are completed properly and have the proper supporting materials. Failure to submit proper attachments can result in immediate dismissal of your application. Even if your application does not require an engineering study, you may want to have a broadcast consultant review your materials before they are submitted. There are no second chances.

Registering with the FCC

In order to conduct any business with the FCC, you will first need to create a username and a password in the CORES system. CORES stands for *Commission Registration System* and is used for all FCC licensing registration filings.

Set up your CORES account as follows:

1. **Visit FCC.gov**
2. **Click "Licensing & Databases (CORES)"**
3. **Click "Register and update Username Account"**
4. **Create username**
5. **Create password**

After you have set up your account, you will be able to log into CORES and create your FCC Registration Number (FRN). FRNs are necessary for individuals, such as consultants doing business through the FCC website, as well as entities such as your nonprofit.

After you receive your 10-digit FRN, you will be able to manage all aspects of your FCC licenses and applications. If you have any trouble registering, you can call the FCC Licensing Hotline on (877) 480-3201 and select Option 4.

When you have your FRN, you need to apply for a Facility ID Number (FIN) for your proposed broadcast station. You can obtain this number through the CORES system after logging into your account.

Collecting the necessary materials

Along with your application (detailed in the next section) you will need to include materials such as the Articles of Incorporation for your nonprofit (see Chapter 2) an engineering study for your proposed tower site (if necessary), a letter of reasonable assurance signed by the property owner, and an educational statement of purpose pledging to use the frequency in an educational way.

Educational statement

Your *educational statement* must provide an understanding of the organization's educational purpose for the FM frequency and an explanation of how a LPFM frequency would help the organization achieve this purpose. Your document does not have to be any exact length but should provide a good overview of your educational plans for the station. It should include the date the organization was incorporated or first recognized by the community. It is important that your educational statement is a truthful reflection of your organizational goals for obtaining an FM broadcast.

A proposed programming schedule should be included with the application. Figure 3-3 shows some programming ideas from WXNA Nashville.

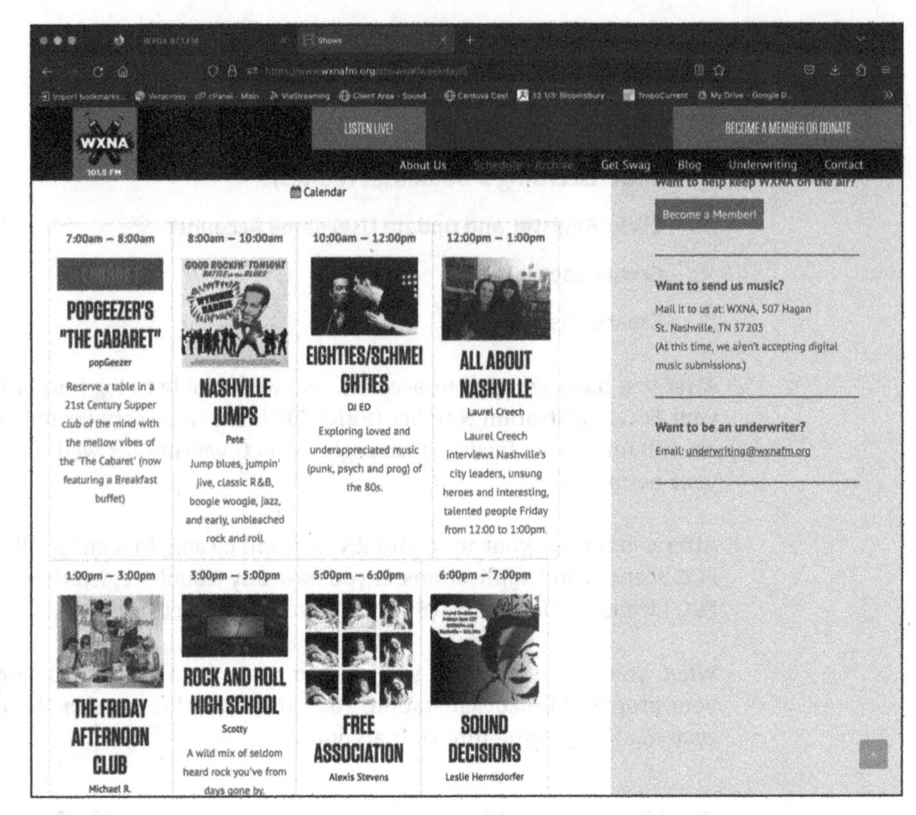

FIGURE 3-3: The schedule on the WXNA website provides some programming ideas.

TIP

The way a station educates its listeners may not be traditional mathematics and ABCs; rather, the station might educate the community about local musicians, female composers, or contemporary artists, for example. Discussions among community leaders and activists could be extremely enlightening to area listeners — as would music programs hosted by teenagers. It's OK to think outside of the boombox!

Letter of reasonable assurance

After you have found your tower site and befriended its generous owner, you need to get a *letter of reasonable assurance* signed by the landlord and submitted to the FCC. This can be a very simple letter that you draft up stating that the owner of the building is willing to let you co-locate your antenna on their property, as stated in your LPFM application, if the license is granted. This letter must include the site location and the owner's contact information, as well as the station name and primary contact. This must be signed by the site owner and submitted to the FCC with your LPFM application. A sample letter of reasonable assurance is provided in the online resources for this book.

TIP

A reasonable assurance letter is not a lease although a signed lease can also work in this situation. However, if your competition is steep, you may want to wait to until your LPFM application is granted prior to getting into a site contract — to make sure that you will have the permission to follow through with the plan.

Keep in mind that reasonable assurance goes beyond just permission from a property owner. You need to verify as diligently as possible that your proposal complies with local ordinances, power availability, historical designations, and neighborhood codes. For example, if you are planning on building a tower in your backyard, does the city or home owners' association allow that, and are there local restrictions on any height? You do not need to have written verification for these things, but it is important to confirm that any proposed construction is locally permissible in the event that your application is challenged by a competitor.

Submitting your application

After you have your username, your FRN, and all your documents in order, you will be ready to file a Form 2100 Schedule 318 "Low Power Construction Permit" on the FCC website (`fcc.gov`) when the LPFM application window opens. The form is pretty straightforward; just be sure to follow the instructions carefully. Application instructions can be found online: `https://www.fcc.gov/sites/default/files/lms-318-construction-permit-application.pdf`.

TIP

The window will be open anywhere from five days to two weeks. Do not wait until the last minute! The Licensing and Management System (LMS) on the FCC website has been known to have problems when traffic is high.

REMEMBER

Before finalizing your application, make sure that you have included all of the required materials. Here's an application checklist for you to follow:

- Articles of Incorporation
- Educational statement of purpose
- Letter of reasonable assurance signed by the tower site's owner
- Engineering study (if required)

Remember, when the filing window ends, it ends. The system will not let you submit the application even if you are 10 seconds past the deadline. You can amend your application later for small things, but you cannot amend engineering errors or the competitive standing of the application.

After triple-checking that the information is correct and complete, take a deep breath and press "Submit." Now, let the games begin!

Winning Your FM Frequency

After the window closes, the FCC releases the names of the organizations that have applied for LPFM frequencies. This moment of truth brings both jubilation and heartache to many organizations across the country. Some organizations learn that they are a *singleton*, which means they are the only applicant for a frequency and a Construction Permit to build their FM station will be sent automatically. Others will learn that they are in an *MX group*, a mutually exclusive group of applicants applying for the same channel. Tentative selectees of the MX groups will be made public by the FCC through Memorandum Opinion and Order (`fcc.gov/documents/memorandum-opinion-and-order`). This information can be received via email by signing up for the FCC Daily Digest (gov/proceedings-actions/daily-digest).

Your possible outcomes are as follows:

- **Win:** If you are a singleton and you completed your application correctly, the FCC grants your organization a construction permit to build an FM station. Congratulations! You may go to Chapter 4 (or read the rest of this chapter to broaden your LPFM knowledge).
- **Lose:** You lose your chance at the frequency if you don't have as many Priority Points the other applicants (see Chapter 2).

  You also lose if you have forgotten to include a critical piece of information — such as your engineering study — or if the FCC finds other flaws or misinformation in your application and you are disqualified. There are no second chances and no one knows when another window will open. You can skip to the end of this chapter for alternative FM opportunities, or flick through to Chapter 5 to find out more about Internet broadcasting.
- **Draw:** If other organizations have applied for the same frequency, and you all have the same number of points, you are in a MX situation. This means that if one application is granted, the others cannot be simultaneously granted.

  If you are in a MX group, you need to immediately size up your competition. You may right away see flaws in their application that will disqualify them from holding the frequency. You may have to dig deeper into their background to find problems with their engineering or organization. You can bring these flaws to the attention of the FCC as discussed in the next section. Or, more collaboratively, you may think they seem like a really cool group that seems like a good partner. Collaborating may be the key to beating out the other competition, and you need a strategy to win. The next section helps you find a way forward.

The FCC does not weigh applications based on the merits of your educational purpose, the mission of your organization, or how much community support you may have. It only requires that you have them. If there is competition for your frequency, they will award the permits based on a points system alone.

Navigating MX Situations

If you find yourself part of a MX situation, you have got to win or find a way out.

One of the easiest solutions to being in a MX group is making a move away from the group and onto a new frequency through a *reengineering escape*. Speak with the FCC or your broadcast engineer to see if there is another open channel in the area that received no applications. If the proposed change is within 11.4 km (approximately 7 miles) in any direction or three channels up or down, a minor engineering amendment can be made at any time. If you wish to move larger distances or over to other open channels, you will need to do so within a FCC-declared *technical change window* (when the FCC is accepting applications for major technical changes).

If you cannot escape your MX group, consider whether you would be interested in partnering with any of the nonprofits in this group. As I discuss in Chapter 2, this can be risky but can also make for a very happy arrangement — and, let's face it, half a frequency is better than no frequency at all!

Winning your frequency

If an MX situation is unavoidable, there are ways to win. But if you are going to play the game, you may have to get ruthless. If you find that an organization has violated any rules or neglected any required information, you may bring it to the attention of the FCC. At any time after the filing window ends you may file an *informal objection* (a letter to the FCC detailing problems with a broadcast station or application) against a competitor. You can only file a *petition to deny* (a formal complaint backed with evidence of wrong-doing presented to the FCC asking them to deny a station application or renewal) against a competitor within 30 days of the FCC announcing the winner (known as the *tentative selectee*) of a LPFM MX group. Through these methods, you can flag an application for review and bring any evidence of violations to the attention of the FCC.

The LCRA requirements are discussed in Chapter 2; a number of ways allow you to investigate the other organizations in your MX group to see if their application has any holes:

- **Missing engineering study:** The first step is to examine their application carefully. Look for any mistakes or forgotten attachments. You would be surprised how often folks forget to include their engineering study or something else just as crucial. If any of the pieces are missing, the application will be dismissed automatically by the FCC. This is why you must be incredibly careful when you file!
- **Character issues:** If a board member has had character issues or has been dishonest with the FCC in previous proceedings, their application will be dismissed.
- **Conflicting ownership interests:** An organization may only file one LPFM application per window and it may not hold any other FM licenses. Board members may not be on the board of any other organization that holds an FM license. Do your research!
- **Failure to meet citizenship requirements:** If fewer than 80 percent of the board members are U.S. citizens, the organization is not permitted to hold an FM frequency.
- **No local presence:** Research the organization, make sure they are who they say they are and that the board members actually know they are applying. Search the Secretary of State database, visit their assumed headquarters, and make sure the organization is truly operating locally in the community.
- **Pirate broadcasting history:** Known pirate broadcasters are prohibited from holding a LPFM license. Any organizations with board members who have participated in unlicensed broadcasting will be dismissed.
- **Previous convictions:** If a board member has had a conviction that has resulted in the loss of Federal benefits under the U.S. Anti-Drug Act, or if they have been convicted of felonies or other crimes involving discrimination or fraudulent statements to the U.S. Government, their organization will be disqualified.
- **Zoning:** If the applicant proposes a new tower, or a tall structure to be built, you should check with the local county or city zoning, or Home Owners Association (HOA), to see if that structure can even be built.

Ex parte rules are in effect when you're investigating other organizations in a MX group. This means that any communication you have with the FCC concerning a competing applicant needs to be copied to the contact representatives for the competing applicants. Competitors that you are filing against must be sent a copy of the pleading via U.S. mail on the same day as you file your communications with the FCC.

If your group is filed against, you have ten days to respond with an *opposition* (your defense against the claims) after you receive the petition in the mail. Speak to your engineer, attorney, or FCC advisor as this can be very serious and you want to respond promptly with the proper information. Pleadings are filed electronically via FCC (go to the LMS login page and look for the "Submit a Pleading" section).

After the FCC has reviewed all the comments and dismissed any applicants that are not qualified, the frequency will be won by the organization — or organizations — in each MX group with the most Priority Points. Chapter 2 presents a breakdown of how these points are scored.

TIP

As up to three organizations within the same MX group may join forces and combine points to beat the other competition, some experts recommend having another friendly organization (or two) apply for the same frequency so that you can collectively win the frequency in a highly competitive situation.

EVERYTHING YOU ALWAYS WANTED TO KNOW ABOUT MX SITUATIONS BUT WERE AFRAID TO ASK

Todd Urick, Technical Director of Common Frequency, has compiled a list tackling the most commonly asked MX questions.

How do you know who is competing for your frequency?

You can obtain a preview of any competing applications by going to "FCC Public Search" and selecting "Radius Search." You achieve this through visiting the website `enterpriseefiling.fcc.gov/dataentry/login.html` and selecting "LMS Public Search," then "Radius Search": (1) Enter your LPFM coordinates; (2) Distance 40 km; (3) Service: LPFM; (4) Channel range -3 to +3 of your channel; (5) Facility Type: All. In the print-out, look for facilities labeled "New." Your primary concern is applications that are on the same channel as you within 24 km of your application.

(continued)

(continued)

When can you make a reengineering escape?

There are two ways to make a reengineering escape: At times when you are MX'd, there may be another channel open nearby that nobody else applied for. In that case: (1) File a minor amendment. You can try to "unlink" yourself from a MX sometimes via a minor engineering modification. This means you can move 11.4 km in any direction or change three channels up or down. This can be applied for at any time. (2) Channel jump. At times, usually within a FCC-declared settlement or technical change window, you can jump to any open channel at larger relocation distances. Check with the FCC or an engineer concerning this option.

What is the best way to beat your competition?

The most common engineering mistake is if an applicant has forgotten to attach a second adjacent waiver request in situations where it is required. You can check. Go to `radio-locator.com` and enter the community of interest. If the channel a competitor applied for has established radio stations on the second-adjacent channel (for example, for 100.1 FM, that's 100.5 and 99.7 FM), and did not include an interference waiver attachment on the application, that could be a defect.

Can I pay an applicant to get them to withdraw their application?

Yes and no. Compensation for dismissal of a competitor's application is limited to money for *prudent expenses* — expenses that went into preparing the competitor's application (consultants, engineering/legal costs of the application, and so on). These costs usually can be justified up to $5,000 without the FCC asking for details.

Joining a timeshare

If there is a tie after the Priority Points are tallied (see Chapter 2 for how points are scored), the competing applications will be encouraged to work out a timeshare agreement within 90 days of the release of the MX group. A *timeshare agreement* is an arrangement where organizations develop a schedule for sharing the frequency.

Organizations that are tied can combine points in order to beat their opponents and win the frequency outright, but no more than three tied groups may combine their points.

REMEMBER

If the groups in an MX situation cannot come to a timeshare agreement, the FCC splits the license equally between them on an eight-year nonrenewable term. The FCC encourages tied organizations to develop their own timeshare agreement and present it to the Commission for their approval. The license could be split on a daily, weekly, or annual schedule.

WARNING

Timeshares may be successful for organizations that share similar missions and can collaborate on programming. It is, however, difficult to brand a station that is only on-air for half a day or half a year. It is also confusing to listeners who may blame you for any questionable programming brought on air by the other organization. Also, although timeshares may be amenable, a number of situations have occurred where one organization eventually tries to sabotage the other — thereby getting control of the entire FM frequency. It is for these reasons that it is important to seek your own FM frequency from the start if at all possible. If not, take what you can get and keep your ears open for new opportunities down the road.

For more on entering into partnerships with other organizations, check out Chapter 2.

Alternative Methods for Getting on the FM Dial

Organizations can apply for new LPFM licenses only when the FCC opens a licensing window, and can broadcast only if they win the frequency. If you miss it or you get beat out, you are out of luck. In order to get the FCC to open another LPFM window your best bet is to speak with your congressional representatives and have them petition the FCC on behalf of your community. As more people are activated on this — or any — issue, the more likely it is the FCC will take action.

Even if you miss out or lose an MX situation, viable alternatives remain. You can work with an organization that already holds a LPFM license or use an alternative method of broadcasting. Part 15 broadcasting is tackled later in this chapter, but if you wish to reach a wider audience, see Chapter 5 and launch your adventure in webcasting!

Working with licensees in your community

If there is no future LPFM window in sight or no available frequencies in your area, you may have to get creative. The first thing to do is check in with existing LPFM stations in your community. Sometimes you can find a license holder who is ready to move onto a different project. They can transfer the license to your nonprofit organization with minimal paperwork and no fees from the FCC. They are allowed to be financially compensated for their time building the station and for any equipment they provide but they cannot make a profit from the transfer of the LPFM license itself. Therefore, if there is any price for the license transfer, it should be a reasonable exchange — unlike acquiring a full-power FM frequency, which is only a possibility for those with very deep pockets.

If there are no opportunities in your community for obtaining your own LPFM license, you may be able to partner with an existing noncommercial station and share the programming responsibilities. For example, an existing high school or college radio station may be interested in providing your group with airtime in the evenings, weekends, and vacation periods when the student programmers are away from campus.

Using Part 15 broadcasting

If all else fails, you can daisy-chain Part 15 transmitters to get legal FM coverage in your designated area. This will not require any permissions, paperwork, or licensing but it will take some work, creativity, and ingenuity.

TECHNICAL STUFF

What's a *Part 15 transmitter*? It refers to Part 15 of the FCC broadcasting rules that permits the unlicensed operation of extremely low-powered FM and AM transmitters. Prior to smartphones, their most common use was for "talking houses" in the real estate world where you could pull up outside a house for sale, tune into a designated channel, and hear about the layout inside. The FCC limits the broadcast range of these devices to approximately a 200-foot radius. Of course, this will depend on the terrain. It could cover a flat football field pretty well. In more dense or rugged areas, maybe not so much. But you can add as many transmitters as you need to cover a larger territory in a daisy-chain.

The magic of radio cannot be diminished, even on a tiny transmitter. During the COVID-19 pandemic, when community gatherings were prohibited, there was a resurgence in interest in Part 15 transmitters. Churches were purchasing them for parking lot services and entertainers were using them for drive-in concerts. I experienced a brilliant production from Kentucky Shakespeare that used Part 15 transmitters for a performance of *MacBeth*. On a dark, foggy night, the audience sat inside their cars listening through the radio to disembodied voices of performers who weaved in and out of the midst toward the stark drama ahead of us on an elevated central stage. I was there, and I was enraptured!

If you wish to experiment with Part 15, you must purchase a transmitter that is legitimately "FCC Certified." If it does not have this stamped on it, or if the wording is different, you may not use it for your broadcasts. In recent years, imposter transmitters that are not in compliance with FCC rules have been appearing online. Using them can result in interference with other broadcast signals and can result in FCC enforcement. Purchasing your Part 15 transmitter from an authorized electronics dealer will help make sure you get the right item. The FCC has started a database where you can find out if a transmitter's authorization number is legitimate; check it out here: `https://www.fcc.gov/oet/ea/fccid`.

WARNING

Beyond Part 15 transmitters, American citizens are not permitted to broadcast on the FM dial without a valid license from the FCC. As excited as you are to get the voices of your community on the airwaves, do not modify your Part 15 transmitter in any way. Getting caught engaging in unlicensed or *pirate* broadcasting can result in massive fines and possible jail time. It also jeopardizes your chances of getting a LPFM license in the future. Don't spoil it for yourself or your organization. Even if you don't get a license in the first round, opportunities arise unexpectedly so you'll you want to stay qualified.

2

Countdown to Blastoff: Launching Your Broadcast

IN THIS PART . . .

Ignite your FM transmission when your application has been approved, set up your broadcast equipment, and organize your launch party!

Broadcast your radio station online and reach out to listeners based both near and far from your studio.

IN THIS CHAPTER

» **Taking the first steps after your application is approved**

» **Assembling and activating your broadcast equipment**

» **Setting up the Emergency Alert System**

» **Hosting your launch party**

Chapter 4
Firing Up Your Frequency

You've done it! Celebrate, rejoice, and make your offerings to Our Lady Radio Mutiny — the dream is officially a reality and now you have to get your station on the FM dial. Building a radio station is no easy task, but if you've got a permit for a frequency, you've already done the impossible. So it's no big deal to move forward from here! Take it one step at a time and soon you will be driving around listening to your brand-new FM station. Nothing is more magical than that! This chapter explains how to bring your new LPFM frequency to life.

WARNING

In this chapter, you'll discover some of the basics of FM installation to help you understand the process and the materials you'll need. However, I strongly recommend that you hire a professional to raise your tower and install your antenna. One wrong move when working with electricity can be catastrophic to the humans and the equipment involved. The information provided here is general guidance rather than a fail-safe manual.

Permission Granted: Getting Started

After the Federal Communications Commission (FCC) approves your application, they issue your organization a Construction Permit (CP), which gives you three years to build a Low Power FM radio station.

Keeping your FCC information safe

REMEMBER

All your future dealings with the FCC use the Commission Registration System (CORES). Your FCC Username and FCC Registration Number, and your Facility ID Number (FIN) are unique identifiers affiliated with your station (more on establishing these numbers in Chapter 3). You will use this information for all your filings within the FCC website, so be sure to keep it in a safe place.

Claiming your call sign

One of the first and most exciting things you can do with your Facility ID number is to apply for your station *call sign* (your official four-letter station identifier). If you are east of the Mississippi River, your call sign will start with a "W." If you are west of the Mississippi, your call sign will begin with a "K." To find out what call signs are available, visit the FCC website Call Sign Search page (`enterpriseefiling.fcc.gov/dataentry/public/tv/publicCallSignSearch.html`) and begin typing four-letter combinations until you get an "Available" result.

TIP

Finding your ideal call sign may take a while. I recommend making a list of every available call sign that you come across. Don't get your hopes up. The list will not be very long. There just aren't that many four-letter combinations left. Still, you might get lucky. Call signs get turned in from time to time and perhaps there are letters that no one else wants that would be perfect for your organization. WHIV, for example, is a call sign that was not in high demand for most applications but was the perfect choice for the New Orleans Society for Infectious Disease Awareness.

TECHNICAL STUFF

United States merchant vessels may retain call signs beginning with "W" and "K." If the call sign you want is reserved in this manner, contact the U.S. government (the FCC will provide the relevant contact details) to ask if that call sign may be reused for your broadcast station. If the call sign you want is used by a TV station, you have the option of contacting that licensee to see if you can reuse the call sign for a LPFM. In many cases, the TV licensee may not permit it, but it is worth a shot.

If there is no obvious choice for your group, you may have to get creative and find a way to make the available letters work. If nothing stands out, talk the choices over with your board members and your friends and family. Think about branding and marketing and which letters seem the most memorable. It is possible to change your call letters in the future but that could require an extensive rebranding process.

REMEMBER

After you have a list of available call letters, move on to your favorite combination right away. After you decide on a set of call letters, reserve them through the FCC's Licensing and Management System (`https://enterpriseefiling.fcc.gov/dataentry/login.html`).

Setting Up Your Signal

Getting your station on air before your permit expires does not mean that your station has to sound perfect or have the programming all worked out. There's still plenty of time to adapt to new technology and meet the needs of your community. At the very least, however, you need to get your signal on the FM dial briefly enough to read out your call letters and test out the transmission.

WARNING

You need to build your station according to the exact specifications of your LPFM application. If you wish to make any changes to the location, equipment, or operation of your FM signal, you have to notify the FCC and provide further engineering information.

Sourcing equipment housing

Every tower site needs a weatherproof shelter with power to install a transmitter. This can be a house, garage, shack, shed, or weatherproof box made for communications equipment. If your remote transmission site does not have a shelter or basic utilities, you need to have them installed.

REMEMBER

The shelter does not need to be complex but does need to be secure. It will house your electronic equipment to protect it from all of the elements and from human interference. In all cases, vents are needed with at least a fan moving heat out of the vicinity of the transmitter.

TIP

If your equipment is housed in an indoor area accessed by individuals not affiliated with your station, you may consider a *locking network cabinet*. Made of a secure metal construction, these cabinets have mesh openings for air circulation on all sides, and possibly built-in fans. Tripp Lite (`tripplite.eaton.com`), NavePoint (`navepoint.com`), and StarTech (`startech.com`) all manufacture these network cabinets for $500 to $600 depending on the size. Be sure to get a swing-out cabinet with removable side and rear panels, as this allows you to access all of the ports and connections.

Piecing everything together

A number of organizations and independent engineers who specialize in LPFM broadcasting will come to town and help you get your station on the air. The Prometheus Radio Project (prometheusradio.org), International Media Action (imarad.io), Brown Broadcast Services (brownbroadcast.com), and Austin Airwaves (austinairwaves.blogspot.com) are legendary community radio wranglers who will do what it takes to get your signal on the air.

In terms of assembling the equipment, you need a technically minded person to assist in putting the system together. Such a person should be able to follow the installation instructions provided by the equipment manuals. You'll also need someone proficient in Internet configuration and someone to ensure that the IT systems are communicating in the correct way. Take it one step at a time, ask for help when you need it, delegate to specialists, and don't be too intimidated. Often the broadcast engineer who helped with your application (see Chapter 3) will be able to help you assemble the technical elements. If they are not there in person, they may be able to walk you through the set-up via a video chat. It may cost extra for this kind of assistance but it will surely be worth it.

TIP

Here's a list of equipment you will need at your FM transmission site. The first three items are essential for all LPFM broadcasters. The audio codec, computer, and IT systems are standard but optional and may not be necessary if your studio is at the transmitter site:

- **Antenna:** Propagates FM waves through space
- **Transmitter:** Generates FM waves from audio input
- **Emergency Alert System:** Provides emergency alerts automatically
- **Computer:** Allows remote system monitoring
- **Audio codec:** Carries studio audio to tower site (if necessary)
- **IT systems:** Provides Internet access
- **FM radio:** Monitors your radio signal

In case you are wondering what you are in for equipment wise, I have included the equipment lists for two transmission sites in the online resources for this book.

Choosing Between On-Site and Remote-Site Transmission

Ideally, your studio site will be in the same location as your FM transmitter so that you can run a line directly from your control room up your tower to your antenna, which is very reliable. This is called *on-site transmission.*

However, many stations have the studio and transmission tower on different properties. They use a Site To Location (STL) link to get the audio to the transmission site. This is known as *remote-site transmission.*

In the simplest sense, an STL can be created using a computer connected to the Internet at the tower site. It uses a web browser to pick up the station's streaming URL and feeds the audio to your FM transmitter.

TECHNICAL STUFF

For more reliable Internet transmission, most studios use a pair of hardware devices called *codecs* that communicate using a dedicated IP address. One of these codecs is placed in the studio and it encodes the audio to travel across the Internet. The other codec, placed at the tower site, then receives the audio from the Internet, decodes the audio and feeds it to the transmitter. Units by Barix (`barix.com`), Comrex (`comrex.com`), and Tieline (`tieline.com`) are just some of the codecs currently sold. It helps to have someone familiar with IP technology to help set up your codec system. The reliability of your codec system depends on the reliability of your Internet provider. If you want more stability for your radio station in times of emergency, you may consider an alternative type of STL link such as a dry-pair telephone line, or analog 950 MHz microwave link. This can be an expensive process to set up but can keep your station on air during times of emergency. More information on setting up different types of STL links can be found in the online resources for this book.

ACCESSING AND SECURING YOUR TRANSMISSION

Having a computer connected at your tower site will save you a lot of trips out into the field. For this you need a basic Internet service and a robust modem. This provides a great way of checking and controlling the equipment at your site remotely through the Internet. Remote desktop software such as LogMeIn (`logmein.com`), AnyDesk (`anydesk.com`), Teamviewer (`teamviewr.com`), and Chrome Remote Desktop (`remotedesktop.google.com`) make this possible.

(continued)

(continued)

It takes some configuration on each piece of the broadcast equipment to make them all accessible remotely, but your equipment manuals will provide step-by-step instructions. After it is all set up, you can connect to your transmitter, EAS system (more on this later in the chapter), and audio codec (detailed earlier in the chapter) directly by logging into this computer.

Remote access allows you to shut down and restart your transmitter remotely. This can come in handy in drastic situations, such as a rogue DJ or a computer hack. The FCC requires that a station transmitter "will be turned off by station personnel within 3 hours of an overpower, overmodulation or out-of-tolerance frequency condition that can cause interference that cannot be corrected within that period of time." Be certain that you can always respond effectively. Also keep a log of any time the transmitter is turned on or off. You need this information in your FCC files.

If there is a way to get a back-up generator at your tower site, absolutely do it! If it is not in your budget yet, plan for it in the future. The power will go out at some inconvenient moment, and in times of emergency it will be really important to stay on the air. Traditional gas-powered generators work well but can be loud and have a strong smell. They need to be refilled and can be a safety issue if not used properly. Solar generators have become a reliable option in recent years and many can be controlled remotely

It is imperative to secure all your systems. Change passwords from the default to something new on all equipment. Sound basic? Well, some LPFM stations failed to do this and found themselves victims to a major hack on Inauguration Day 2017 because none of them had changed the default password on their audio codec devices.

Understanding Your FM Transmitter

Your *FM transmitter* is the device that generates the FM signal.

REMEMBER

LPFM transmitters must be FCC Type Certified and marked accordingly on the shell of the equipment. Be sure to confirm that this is the case before purchasing; otherwise, you will be left with a perfectly good piece of equipment that you will never be able to use. Companies that build FCC certified transmitters include GatesAir (`www.gatesair.com`), Nautel (`nautel.com`), NiCOM (`www.nicomusa.com`), and Ramsey Electronics (`ramseytest.com`). Expect to spend between $3,000 and $6,000 on your FM transmitter. It will be one of your larger, more important purchases.

Many transmitters allow you to simply plug in your audio, antenna, and ethernet cable for remote monitoring. Ask your radio vendor which models have built-in processors, dynamic RDS, and web-based monitoring control. Don't purchase a cheaper transmitter that requires expensive additions to be satisfactory. In the online resources for this book, you can find more information on transmitter details and terminology. Please check it out if you wish to learn more about the different transmitter features.

A favorite option that now comes standard in many machines is a *Radio Data System* (RDS), which is a protocol for displaying dynamic text via FM subcarrier data. This allows text to show up on FM receivers with digital displays (including those on most vehicle dashboards). This is a fun way to send short messages or song information to listeners. Do not fret if your transmitter cannot do this, as standalone boxes can be added to your broadcast chain to permit RDS.

TECHNICAL STUFF

In most cases, the LPFM license permits the generation of 100 watts of Effective Radiated Power (ERP) at 100-feet antenna height above average terrain (HAAT). The power gets adjusted by the FCC to maintain the same broadcast range. The higher the antenna, the lower the wattage, and vice versa. Your ERP, which is the amount of power your antenna is pushing out of the antenna, is different from your Total Power Output (TPO) exiting from your transmitter. To account for line loss as the energy moves up the cable as well as gain on the antenna, you may need a transmitter that can produce more wattage than you will radiate. So, if you will be broadcasting 100 watts of ERP, you will still need to purchase a 200- to 300-watt transmitter. The formula is:

(TPO – line loss in watts) × (Gain of antenna in times gain) = ERP

An example of this formula in action is included in this book's online resources.

Your ERP and TPO will have been calculated in the engineering information you presented to the FCC in your LPFM application (see Chapter 3). You must adhere to these settings when installing your transmitter.

WARNING

Never turn on your transmitter without a license. Unlicensed use of FM transmitters can result in massive fines, loss of LPFM privileges, and possible time in jail! To keep yourself out of trouble, make sure that any time your transmitter is on, it is used exactly according to the terms of your license.

Never turn on a transmitter without an antenna connected. Doing so has the potential to severely damage the transmitter.

Installing Your Antenna

Your *antenna* is the piece of equipment that works in conjunction with the transmitter to propagate radio waves through space.

LPFM antenna choices can range from the inexpensive Comet CFM9557 (`cometantenna.com`) often used in amateur broadcasting, to the professional grade Jampro JLLP-11 (`jampro.com`) or the Nicom BKG77 (`nicomusa.com`).

WARNING

To hang your antenna, your shining piece of jewelry in the sky, I recommend hiring a professional. This is a highly technical, often dangerous job, and I do not recommend tower climbing for the inexperienced! Just don't do it. Perhaps your broadcast engineer or some of the professionals I recommend earlier in this chapter can come to town to help you install your antenna. Telecommunications companies in your area can put you in touch with local tower climbers (though it is unlikely they will have experience hanging FM antennas to get maximum coverage). Whoever they are, tower climbers need to be insured during installation; this insurance can be purchased by the day.

TECHNICAL STUFF

For those who wish to dig deeper into the antenna installation process, please refer to the online materials of this book for information prepared by Todd Urick of Common Frequency (`commonfrequency.org`) with assistance from Paul Bame of the Prometheus Radio Project (`prometheusradio.org`).

If you do plan on hanging the antenna yourself, please keep in mind that working with electricity can be dangerous. FM broadcast installations must be properly grounded. *Grounding* is a process of transferring lightning and static from the tower and antenna to the earth instead of it penetrating your broadcast chain and studio via the studio cable feedline.

To install your antenna, you need

- A coax ground kit that matches the type of cable you are using
- A ground rod
- AWG #4 or #6 copper wire to connect the antenna to the ground rod
- A ground clamp (for the wire to attach to the ground rod)
- A coax surge protector

After you're fully equipped, the process of installation is as follows:

1. Establish a reliable way for the ground wire to attach to the base (such as a Rohn tower base ground clamp or similar).

2. Install the ground rod at the base of the tower, monopole, and mast. Attach this to the tower/mast with copper wire and clamps.
3. Install the coax ground kit to the coax at base of tower and connect to the ground rod.
4. Install the coax surge protector at the location where the coax is going into the building. This should be grounded to the structure.
5. Follow the grounding instructions included in this book's online resources.

Figure 4-1 illustrates two different grounding procedures for mounting an antenna on a free-standing tower and on a pre-existing structure.

TIP

A solid reference is the U.S. National Electrical Code (NEC) sections 800 and 810 (`www.esfi.org`). Keep in mind that every antenna installation is different, so there is no one-size-fits-all method; don't be afraid to reach out to equipment manufacturers if you have questions about their products.

WARNING

It should be understood that even the most comprehensive grounding can still yield extensive loss to transmission systems, and incorrectly grounded electrics are extremely dangerous. Therefore, and once again, I recommend always consulting a professional for your grounding work.

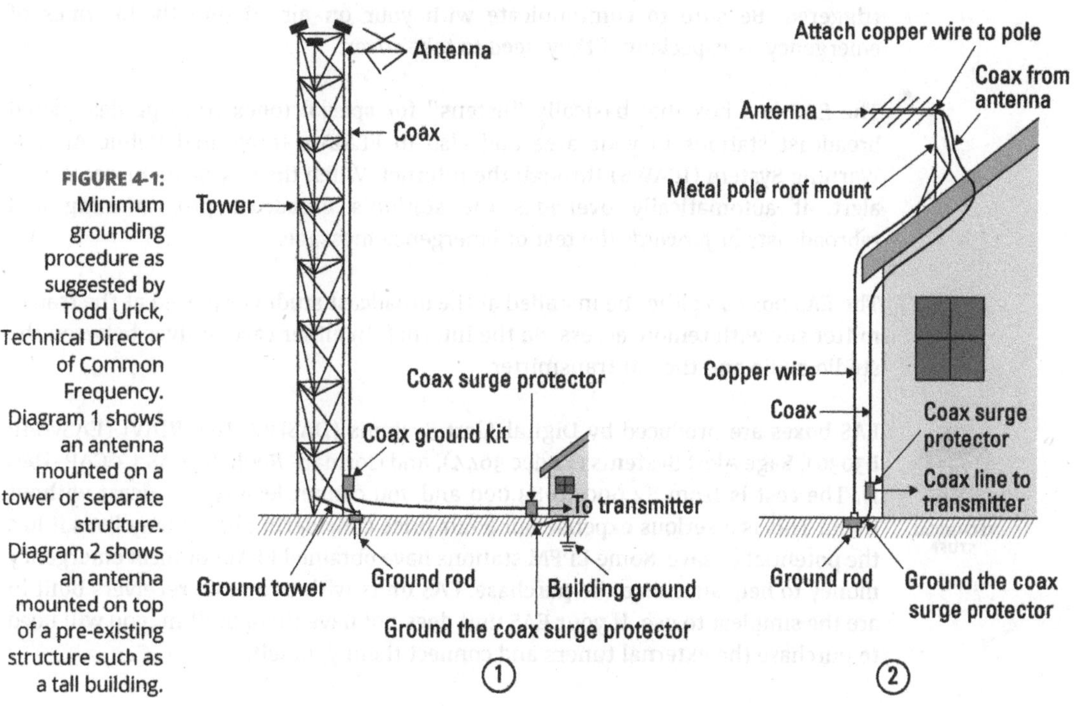

FIGURE 4-1: Minimum grounding procedure as suggested by Todd Urick, Technical Director of Common Frequency. Diagram 1 shows an antenna mounted on a tower or separate structure. Diagram 2 shows an antenna mounted on top of a pre-existing structure such as a tall building.

Installing the Emergency Alert System

FM broadcasters are required by the FCC to have an *Emergency Alert System* (EAS) installed at their transmission site. The EAS is a national warning system designed to deliver critical information to American citizens as quickly and effectively as possible. Stations may deliver local alerts on a voluntary basis but provide the capacity for the president to address the public during a national emergency.

The Federal Emergency Management Agency (FEMA), the FCC, and the National Weather Service (NWS) work together to maintain the EAS and provide pertinent local alerts. FEMA is responsible for issuing any national-level activation and tests of the EAS. Alerts may be generated by federal, state, and local authorities but not the FCC itself. You can learn more about the EAS system here: `www.fcc.gov/emergency-alert-system`.

In times of danger, the EAS automatically interrupts the station's audio and replaces it with emergency information. When you set up your EAS box, you have to select which types of emergency warnings are relevant to your community. Some, such as nuclear bomb alerts, are required nationwide. Others, such as volcano eruption warnings, are not necessary for most states but imperative in Hawaii. The EAS also generates required weekly and monthly tests. Unless the DJ is listening on the FM signal, they may not even be aware that the signal has been triggered. Be sure to communicate with your on-air DJ directly in times of emergency — especially if they need to take cover!

The EAS is a box that basically "listens" for special tones from predesignated broadcast stations in your area and also to FEMA's Integrated Public Alert & Warning System (IPAWS) through the Internet. When the box hears or *decodes* an alert, it automatically overrides the station's broadcast programming and rebroadcasts or *forwards* the test or emergency message.

The EAS box can either be installed at the broadcast studio or placed at the transmitter site with remote access via the Internet. In either case, it lives between the studio audio and the FM transmitter.

EAS boxes are produced by Digital Alert Systems (DASDEC III), VIAVI (EASyCAP B3030), Sage Alert Systems (Endec 3644), and Gorman-Redlich (EAS 1 + CAP-DEC 1). The cost is from $2,000 to $3,000 and you cannot legally broadcast without them. This is a serious expense but a fair price considering how many lives it has the potential to save. Some LPFM stations have obtained FEMA or local emergency money to help subsidize the purchase. EAS units with two radio receivers built in are the simplest to use. If your EAS unit does not have them built in, you will need to purchase the external tuners and connect them yourself.

Each area of a state has a local EAS designation plan which tells you which frequencies to monitor. You can find this information by contacting the representative of the State Emergency Communications Committees found on the FCC website (fcc.gov/SECC-Resources).

Be sure that your EAS box decodes a weekly test from each of the sources it monitors. Your station is required to forward the EAS test script at least once a week. Required Monthly Tests (RMT) are conducted by the Emergency Communications Committee for each state. If your EAS equipment is set properly, all of this happens automatically. Each year a nationwide test is conducted by the FCC. Stations are required to file a Form One through the EAS Test Reporting System (ETRS; www.fcc.gov/general/eas-test-reporting-system) before this test is conducted, letting the FCC know what sort of equipment you are using. Two more forms need to be filed after the test is complete, confirming that you decoded and forwarded the test.

You must keep records of every time your EAS system is activated by a test or an emergency going back at least two years. The box should be set to send confirmation emails to your designated address whenever an alert is decoded or forwarded, and it's a good idea to print out these notifications and keep the hard copies onsite. If your EAS system fails or needs repair, you must note this in your log, too. You have 60 days to get it up and running again.

Be sure to keep your EAS box operating with the latest software with all the current updates. Human lives in your listening area depend on it! You must be ready to protect your community when the time arrives. In fact, you should connect with the technicians at your EAS equipment's manufacturer to review your settings and make sure that your equipment is set to function properly. They can log in remotely and review your set-up. This gives you and your community true peace of mind.

By law you need to get a copy of the *State Emergency Alert Plan* and the *FCC EAS Operating Handbook* to keep at your transmitter location. You can find copies of both on the FCC website (fcc.gov/eas). You can also get a copy of the *State EAS Plan* from your local Department of Emergency Management. Any questions about the EAS rules may be addressed to the Commission's EAS office by phone (202-418-1228) or by email (eas@fcc.gov).

Lighting It Up!

Although your engineering study will be able to estimate the range of your transmission (see Chapter 3), even science cannot predict exactly how the signal will sound to the human ear when the transmitter is fired up, releasing its radio waves into the atmosphere. Radio is eternally fascinating as the signal changes with the terrain, the traffic, and the troposphere (the what? Check out Chapter 13).

The FCC requires that you turn on your FM transmitter and test the broadcast signal before they will transform your Construction Permit into a LPFM License. So go ahead . . . what are you waiting for?

Power up your transmitter to the wattage allowed by your engineering study and see where it goes! You don't have to tell anyone just yet, or you can, totally up to you. At ART FM we decided to do a few test runs before going official at our big celebration. It was fun to see who caught us while we were sneaking on air, playing a few songs.

When you are satisfied that your on-air signal is compliant with the terms of your LPFM Construction Permit, you need to report this information to the FCC on the FCC Licensing and Management System (LMS; `https://enterprisefiling.fcc.gov/dataentry/login.html`) using Form 319. Filing instructions for completing the form are found on the FCC website: `https://transition.fcc.gov/Forms/Form319/319Fill.pdf`.

After it is submitted, the Commission reviews this information and if they find it satisfactory, they grant your License to Cover (your LPFM license). You may continue broadcasting while you wait for this to be finalized.

Keeping up with the FCC

When your LPFM license arrives, frame it! No, really, it is a beauty that you should be proud of, so be sure to find it the perfect place.

At this point, most of your paperwork with the FCC is complete but you will need to stay compliant. The FCC requires minimal reporting from LPFM stations but some things are critical. The station must keep up with EAS tests, transmitter logs, political files, and records of station-conducted contests (outlined earlier in this chapter). Chapter 10 covers the federal broadcasting laws that keep your broadcasters on air and within their legal bounds.

The FCC provides a LPFM self-inspection checklist on their website (`fcc.gov/document/fm-low-power-station-self-inspection-checklist`).

Renewing your license

Your LPFM license will normally expire after eight years. The expiration date is written on your license (that you've framed and have hanging in your station, right?). Your renewal application Schedule 303-S must be filed at least four months ahead of the license expiration, but you are able to apply up to six months in advance. Waiting until the expiration day may incur a hefty fine on your organization.

You need to submit your information through the FCC's Licensing Management System website (`https://enterprisefiling.fcc.gov/dataentry/login.html`). There are no filing fees, but if you fail to renew, you will lose your license.

Within five days of submitting your renewal, you must start making post-filing announcements on air. These announcements use text provided by the FCC and need to be made over a four-week period on weekdays between 7 a.m. and 11 p.m. You can find the text the FCC expects you to read on their website (`www.fcc.gov/media/radio/broadcast-radio-license-renewal#postfiling`).

Stations that are silent during the renewal period still need to renew and post the public notice in a conspicuous spot on their website for 30 consecutive days. If you do not have a website, the filings must be made on a publicly accessible website such as a digital bulletin board run by the city government or a local newspaper. Be sure to keep a record of when and where your post-filing announcements are made.

Keeping the light on

After you launch, you are required to stay on air. Any time you turn your transmitter on or off, you need to note it in your log. If your station will be silent for more than 30 days, you need to request a Special Temporary Authority (STA) from the FCC from the LMS system which allows you to remain silent for up to six months. If that is still not enough time to get your station up and running again, you can file another six-month extension of your STA. You need to notify the FCC immediately when you get your signal back on air. After one year of silence, your license automatically expires and reverts back to the FCC (see Chapter 3 for more on the application process).

If the time comes that you feel like you can't manage to get your station back on air, it is easy to file a Schedule 314 to transfer control of the license to another nonprofit organization in your community. There is no fee for this. Although you cannot sell your frequency to another nonprofit, you can have them reimburse you for the expenses involved in acquiring it (but don't get too greedy or it will raise FCC eyebrows). The new nonprofit will have to get the station powered up, at least long enough to announce their call letters and test the frequency, before the STA expires. Then, they can apply for another extension.

Whatever you do, after you get awarded a LPFM frequency, keep it in the hands of the people!

Partying at Your Official FM Launch

Launching on the FM dial is the very best part of setting up your radio station! Get ready to have a big party when you officially launch your FM signal. This is a big deal — to you and to everyone else. Prior to this day, of course, you will have tested your FM transmitter with secret experimental broadcasts so that you can feel confident that everything will go right.

The official FM launch should be a very big community event (see Figure 4-2). Notify the press and fire up the band! Plan on an exciting countdown. It's not often that a new FM station goes on the air, so make it an event to remember forever.

When you are live, it is always fun to drive around and listen to the extent of your transmission. I think you will be pleasantly surprised! The reach of your signal may change depending upon weather, interference, tropospheric ducting, and the type of car you are driving — but embrace the moment, feel the power!

FIGURE 4-2: WXOX 97.1 FM launch party on Valentine's Day 2016. Bonny "Prince" Billy, Carly Johnson, and the Hundred Proof Stringband were the first live performers on the FM broadcast.

© Aron Conaway, WXOX / Panda Fox Media.

IN THIS CHAPTER

» Launching your radio station online

» Putting the right technology in place

» Helping listeners find your station

» Developing player apps and skills

Chapter 5
Weaving Your Webcast

We live in such an exciting time in which broadcasting isn't limited to the radio! Audiences aren't limited to their immediate community, and your Low Power FM radio station isn't limited to the FM dial. Stations now have the ability to stream their audio over the Internet and reach listeners anywhere in the well-webbed world.

Online broadcasting, *audio streaming*, *webcasting*, *web streaming*, or just plain old *streaming* are terms that are used interchangeably for this type of broadcast. Every day, listeners tune in to the ART FM web stream from distant parts of the world. It's amazing to think about where they are, what they are seeing, and what is happening around them as they listen. If connecting with international audiences and bringing them homegrown programming from your part of the world sounds like an inspiring project, this chapter helps you make it happen!

REMEMBER

Every FM station should also have a web stream. I don't think there can be much doubt about this anymore. It isn't a choice of one or the other; the only answer is both. FM is resilient, permeable, and democratic. Having a terrestrial license will make your station discoverable and powerful within your community (find out more about gaining your license in Chapter 3). Being one of a limited number of FM broadcast stations in your area will automatically give your organization discoverability and a sense of importance that webcast-only stations often struggle to find. However, the ability to connect with online listeners all over the world in real time is an astonishing development that broadcasters have been dreaming about ever since the dawn of radio. Embrace it! What a feeling when you find out you have listeners not only in Canada and Mexico but also in Italy, Morocco, Korea,

Peru, Kenya, Indonesia, Australia, and more! Add in those moments when you realize listeners from warring countries are tuning in at the same time, and suddenly you can imagine your broadcast — the words you are saying, the music you are sharing — as the key to world peace and the secret for happiness around the globe.

When we first started ART FM as a streaming station in 2012, people laughed at the idea: "Why waste your time broadcasting online? . . . Nobody will listen on their computers." Yet when we did, listeners found it so cool to be able to tune in to our broadcast from all over the world! I guess folks got used to the whole thing because ten years later, in 2022, we were hearing the exact opposite: "Why bother broadcasting on FM . . . Everyone is listening online."

The truth may not be as polarizing as it seems. In fact, it is possible we are living in the perfect balance of FM radio and online listening. The FM broadcast reaches our community — all of our community — through its free and readily available transmission. In turn, the webcast allows our community to reach the hearts and minds of listeners all around the world.

REMEMBER

The means for connecting with our audience may be shifting but the purpose remains the same. Radio stations should embrace the FM broadcast and the webcast as different, but equally effective, means for reaching wider audiences. If I had to choose one over the other, I'm not sure which way I would go. The great thing — and the magic of living right now — is that we don't have to make a choice. We can have it all!

In this chapter, you'll discover how to establish an online audio stream as a supplement to your FM transmission, helping you connect with listeners worldwide. Although it will involve additional expenses, including music licensing fees (see Chapter 9 for more on expenses), in today's digital landscape your station simply can't afford *not* to broadcast online.

DIVING IN HEADFIRST

As you can tell by our name, ART FM never meant to be an online station. Our sights were always set on the FM dial. When we first incorporated, the LPFM application window was still a few years off. We used the time to meet with local leaders to gauge their interest in an arts-focused station. One day over coffee, Kris Kimel, founder of IdeaFest, said: "Why wait until you get the FM license? Why not start your station online now?"

He continued: "Why not bring it to IdeaFest? You can set up in the lobby of the Kentucky Center, broadcast from there, and interview some of the guests."

"Okay!" is all I remember saying. We shook hands and he was out the door. I was left feeling so excited about the opportunity, imaging everything we would have to do to make it happen. Suddenly, it dawned on me; I didn't know when IdeaFest was.

"Six weeks!" I exclaimed when I looked up the website back home. "Six weeks to build a radio station! No equipment, no money, no volunteers — all we have is an idea!"

"You can do it," is all Sean needed to say. And together we did.

As luck would have it, the next Grassroots Radio Conference (GRC), an annual gathering of community radio stations, was just a couple of weeks away. I knew I'd find the advice I needed there . . . (Cut forward to the GRC) . . .

"Okay, let me see your notebook," Norm Stockwell, then Chief Engineer at WORT Madison, said in the middle of a large assembly room at the Champaign Urbana Independent Media Center where that year's conference was hosted. He took my pencil and proceeded to sketch a picture of exactly what we needed for an online set-up. This was the first time I ever *saw* ART FM.

Upon returning home, the equipment was ordered and word was spread all over town: "New radio station looking for DJs." We held community meetings and invited interested parties to apply for timeslots. After an article in the local newspaper, the response was quite overwhelming.

Even though we knew we would be streaming online for the first few years and not beholden to FCC rules, we wanted to provide a legitimate broadcaster training program from the very beginning.

To prepare for IdeaFest we began our community training with federal rules and moved into equipment training when everything arrived. Sean set up the sound studio in our dining room and we invited the community inside to learn how to use it. Anyone and everyone was welcome. I suppose we were a little naive.

Word got out that we were looking for technical assistance and Brian McMahan stepped up. Brian is the singer and guitarist of Slint, a band with a very distinct sound. I was a bit startled to hear his familiar voice in calling me on the phone. "Yes," I said, "we definitely need volunteers."

From there, Brian and Sean worked together to get the stream going and the audio equipment sounding just right. I worked alongside ART FM high school intern Alexus Andretti to get the schedule completed and all the materials assembled. We began pairing community members with the visiting IdeaFest speakers, prepping them for interviews. There wasn't much practice time before we launched but somehow we made it happen!

(continued)

(continued)

On September 19, 2012, from IdeaFest in the lobby of the Kentucky Center for the Performing Arts, surrounded by family and friends, ART FM officially took flight! Suddenly, through the wonders of webcasting, we were connecting with listeners all over the world! Listeners were all tuning in to hear our interviews with the world-class guests IdeaFest brought to Louisville — and they were singing our praises on social media. Our first interview was with National Book Award winner Nikky Finney, and our first sound drop was from Grammy award nominee Janelle Monáe. The international listeners were coming in . . . Norway, Japan, Madagascar . . . we were immediately hooked! We loved the international attention!

We kept the stream going for the duration of IdeaFest and then we shopped our online broadcast around to any event that would have us. Couch surfing radio-style, we called it. This got us all kinds of broadcasting opportunities from Matmos to the Kentucky Opera.

In April 2013, we were given a stable home and we have kept our stream going continuously ever since. Ten years feels like a huge accomplishment. Three of those years were prior to the launch of ART FM's terrestrial signal on WXOX 97.1 FM, which gave us an opportunity to build our audience online.

My first love is FM broadcasting but there is nothing like being able to send songs and dedications out to friends and family who live far away! Over the COVID-19 pandemic we developed methods of interactive broadcasting that allows them to send their voices right back! I am certain that we have just dipped our toes into the vast ocean of online broadcasting. The water's deep and I encourage you to dive headfirst into the journey!

Launching Your Station on the Infinite Dial

It is not difficult or expensive to set up an online station. Take it one step at time and you will be worldwide before you know it!

TIP

Here's a list of the basic steps you need to take to set up an online station (the first four actions are covered in more detail in the next few sections):

1. **Connect with a stream host to establish your service.**
2. **Get a direct URL from the stream host.**
3. **Embed a button on your station's website that plays your direct URL.**
4. **Add the studio audio through a computer with audio encoding software.**
5. **Listen online to make sure that your station is broadcasting!**

Connecting with a stream host

A *stream* is a digital broadcast that listeners can tune into online. The first thing you need to do to establish your stream is to connect with a provider who hosts your audio stream in the *cloud* (on their own Internet server, where it can be accessed by listeners connecting online). After you are set up, you use encoding software to send your studio audio to a host who keeps the party going at a designated address in the cloud. Listeners then connect to it from their phones or computers and enjoy!

TIP

You can find a suitable provider by shopping around. Stream hosting should not be too much of an expense. The factors you need to consider include the *bit rate* (the amount of data that is transferred within a certain amount of time) and the number of listeners that can connect at any one moment. Put simply, the higher the bit rate, the richer the audio will sound. Although some services offer inexpensive services of 64 kbps, there is a noticeable difference in audio quality. I recommend using at least 128 kbps. At the time of publishing, it is difficult to find a hosting service that offers audio quality higher than 128 kbps. The few that are out there are very expensive. Perhaps this situation will improve in years to come. The number of allotted listeners can vary greatly between hosting services, so be sure to check the specs. Most likely, a new LPFM station won't need more than 50 slots in the beginning, but you want to be in a good position to grow without changing providers in the future. It is possible to find a host that offers up to 1,000 simultaneous listeners at 128 kbps for around $15 per month. Stream host services include Soundstack (`soundstack.com`), Streamguys (`streamguys.com`), and Broadcast Tool & Die (`broadcasttool.com`). The Pacifica Affiliates Network also offers stream hosting as part of its technology package (`pacificanetwork.org/pacificainternetpackage`).

Getting a direct URL

After you have established a host provider and service, you need to get a *direct URL* from the stream host. This is a dedicated web address for your station audio stream that opens automatically when you visit the site. It does not need a third-party player application. Sometimes your direct URL is provided by the host, but other times you must request it specifically.

This URL will be needed for programming player buttons and station automation systems. For example, the direct URL for the ART FM stream is `https://patmos.cdnstream.com/proxy/artfmin1/?mp=/stream`. If you type this URL into a web address bar, it begins playing ART FM automatically.

REMEMBER

When your direct URL is in place, you need to add a web player button to your website so that listeners can find it.

Embedding player buttons on your website

REMEMBER

If your station's website has one purpose, it is turning visitors into listeners by getting them to hit the play button! Make it easy and obvious for your visitors to begin hearing your audio.

The *web player*, an icon on your site that listeners press to play your online stream, should be immediately noticeable when guests visit your website. Everything on your website should encourage visitors to push play! In fact, they should not be able to resist and there should be no obstacles in the way. In Figure 5-1, you can see how Hollow Earth Radio KHUH from Seattle does a great job at placing the web player button in the front and center of their website.

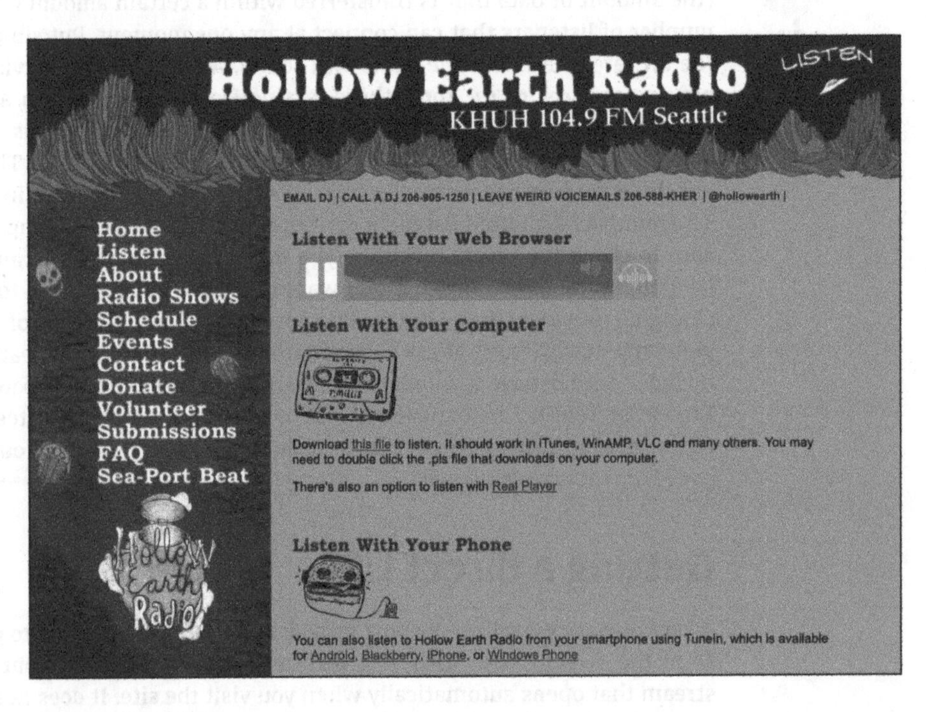

FIGURE 5-1: Hollow Earth Radio KHUH 104.9 FM in Seattle uses a player embedded in their website, `hollowearthradio.org`. They also provide links to other ways to listen.

TECHNICAL STUFF

At ART FM, we presently have two player buttons on our station's home page (see Chapter 2). A button designed by our stream provider is placed in the upper lefthand corner of each page of our site. When browsers make security updates, however, we have had some issues with the functionality of this button. Therefore, we added another player button right in the middle of the home page using basic HTML code that will reliably work on any web browser. The script is provided below; you can use this code to build a player for your own website.

Copy the code exactly as it is shown, and simply replace the sample stream URL (the text "`https://sample.sampletream.com/stream`") with your own stream URL. Then paste the player code into the body of your website.

```
<!-- START OF THE RADIO HTML5 PLAYER EMBEDDING --> <audio
   controls="true" preload="none"><source src=" https://sample.
   sampletream.com/stream" type="audio/mp3">Your browser does not
   support the audio element.</audio> <!-- END OF THE RADIO HTML5
   PLAYER EMBEDDING -->
```

Utilizing pop-up players

Rather than having your player button embedded on your home page (as described in the previous section), you may prefer to have a player that pops open in a new, smaller window. A pop-up player is preferable for stations that don't want their listeners navigating away from their audio stream while searching the web. The pop-up player for KPSQ, shown in Figure 5-2, includes song information as well as alternative links for accessing the livestream.

TIP

You can install a pop-up player by adding the correct HTML code to your website. If you're new to web design, visit W3 Webschools (`https://www.w3schools.com`) to find helpful code script and tutorials.

Adding the audio element

After you have established your host stream and developed your player buttons (well done on achieving these major steps towards streaming your station!), you need a way to bring in the audio emanating from your studio.

The most straightforward way to make this possible is by using a studio soundboard that has a USB output (equipment choices are discussed further in Chapter 6). Connect an ethernet cable from this USB output and feed it into a USB input on the computer that you have dedicated to generating your online stream.

TIP

Your streaming computer cannot be the same computer you use to feed audio to your soundboard. It needs to have Internet broadcasting software installed. Free options for PC computers include Rocket Broadcaster (`rocketbroadcaster.com`) and a program with the terrible acronym B.U.T.T. — Broadcast Using This Tool (danielnoethen.de/butt/). For Mac computers, Audio Hijack (`rogueamoeba.com/audiohijack`) is offered through a one-time licensing fee of $60 and can be used on all of your computers. All of these are user-friendly programs, and you should be able to set up any of them quickly enough.

FIGURE 5-2: KPSQ in Fayetteville, Arkansas, uses a pop-up player that includes information about current and upcoming shows.

Your stream host provider will provide the specific information necessary for setting up your broadcasting software. This will include the server type (Icecast or Shoutcast); the host name/IP address; the port number; the mountpoint; the username and initial password; and the encoding quality. Enter this information on the settings page of your broadcasting software. After you get it in there, be sure to save it as a preset so that you don't have to do it all over again next time. Figures 5-3 and 5-4 demonstrate entering the stream access codes in the Audio Hijack and Rocket Broadcaster software.

WARNING

Pay attention to the bit rate of the encoding format and make sure that it matches your stream host's bit rate (likely 128 kbps). If your host provider places a limit on bit rates, and you go over that bit rate, your stream server will crash. The stream will remain down until you or another administrator logs in to restart the server.

After the codes are entered and everything is set up correctly, simply push the button on your encoding software that starts your broadcast. This is usually a red button or it could be a button that literally says "Start Broadcast." If you get an error message, check all your settings carefully. If you continue to have problems, check Chapter 13 on Murphy's Law and troubleshooting advice.

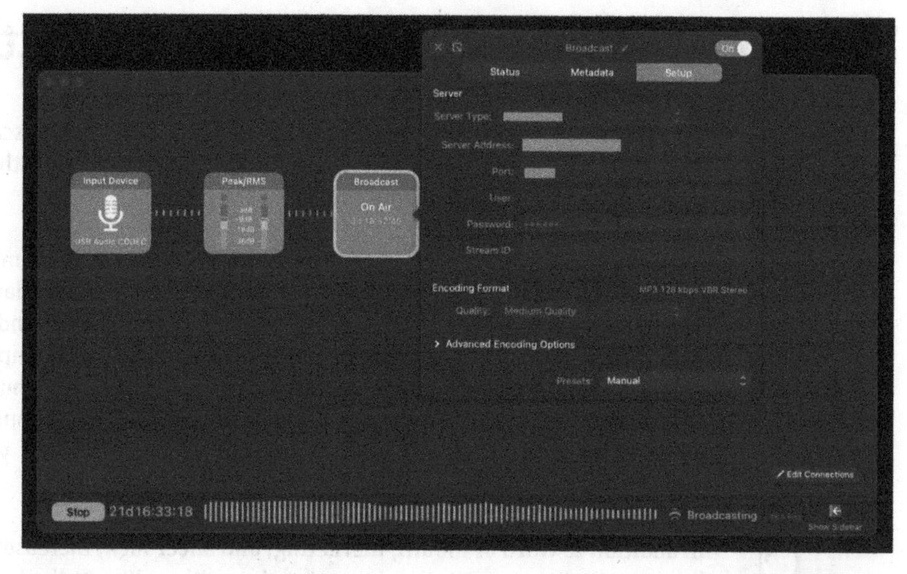

FIGURE 5-3: Broadcast blocks in Audio Hijack for Mac users must be filled in with the codes received from the stream host.

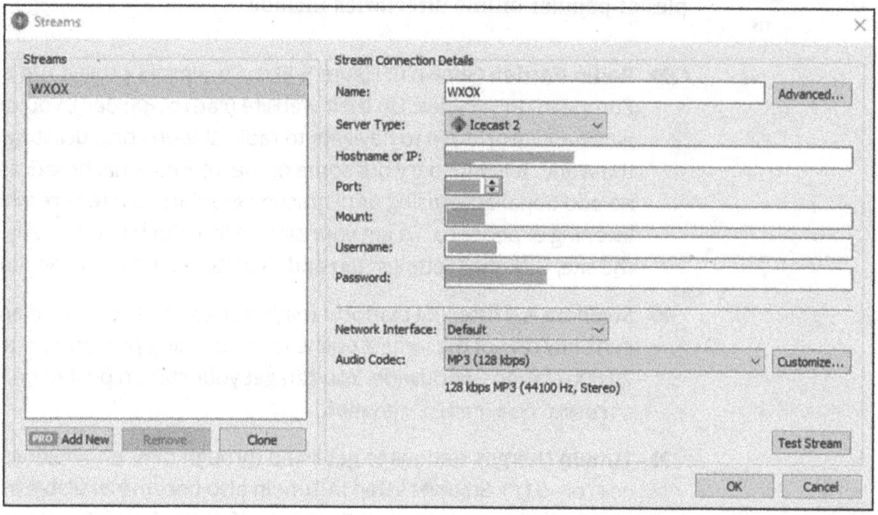

FIGURE 5-4: Stream connection details in Rocket Broadcaster for Windows users.

If the levels start moving on your encoding software and the counter starts ticking — congratulations, you have done it! You are on air! Listeners can now visit your website, push the player button, and enjoy your broadcast anywhere they may go in this well-webbed world!

Registering with Online Radio Directories

Unlike the FM dial where your broadcast can be discovered by scanning through the channels, the likelihood of a listener finding your station in the infinite sea of online radio is low but not impossible.

More often than I expect, we get fan mail from listeners who stumbled across ART FM on some online listening platform or another. The key is to plant a lot of seeds. Make sure your web stream is listed in every directory possible and that your website is printed on all of your material. Have flashy cards printed up with a scannable QR code that you can leave like bread crumbs everywhere you go. Hand them out to new friends as you travel throughout the world. Think about your station as bigger than just a local resource — it is an audio portal into your part of the universe.

TIP

In addition to word of mouth, marketing, and sheer luck, increase your chances of listeners finding your station by registering with online radio directories. Examples of popular online directories include

- **Radio Garden** (shown in Figure 5-5) is a nonprofit project run by a coalition of European universities. On their website (`radio.garden`), you can spin a virtual globe and drop a pin to navigate to radio stations broadcasting anywhere in the world. It is fun to try out some of the international broadcasts! The site has no additional advertising or distracting graphics to interfere with the pure listening experience. To get your station listed in Radio Garden, simply visit the site, click the "Settings" tap and then "Submit Your Radio Station."
- **Streema** is a listening platform that's free to use and easy to enlist. They are currently one of the largest online radio listening platforms with more than 70,000 stations worldwide. You can get your station posted on Streema at `streema.com/radio_request`.
- **TuneIn** charges stations to get listed through their on-air service (`tunein.com/on-air`). Stations listed in TuneIn also become available in Google Home and Apple Music. Legacy stations grandfathered in to TuneIn can update their listing for free at `https://help.tunein.com/contact/update-station-HyB_TSOf`.

New streaming radio directories and applications are popping up all the time. Keep your ear to the ground and make sure that your station makes a splash on them all!

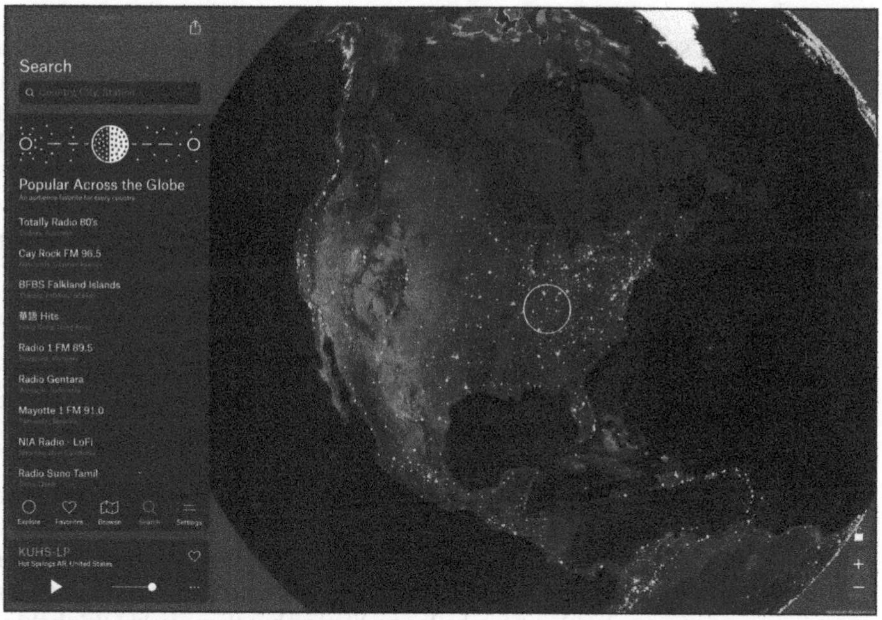

FIGURE 5-5: The Radio Garden website (radio.garden) allows listeners to navigate to radio stations all around the globe.

A plethora of other directories, such as Triode, RadioMobi, StreamTuner, Rhythmbox, and Nextcloud get their information from the Radio-Browser Database. If you would like to see your station included in these listings and reach a larger audience, submit your information at `www.radio-browser.info/add`. Adding your listing is easy enough to do yourself by using your streaming URL. The service also requests your longitude and latitude so that they can geo-locate your station on a map. This only needs to be approximate.

Although the Radio-Browser Database is a free, open-source project, be aware that many third-party listening applications that draw from your stream from this resource will insert visual or audible advertisements or find other ways to monetize your stream. Still, these players may be a way to connect with a wider audience — especially if you use on-air announcements to encourage those listeners to connect with your station directly through downloading your app or visiting your site online.

Developing Apps for Listeners on The Go

If you want to stay connected with your listeners everywhere they go, a station app is the very best way to do it. Short for *application*, an app is a program that is downloaded to a mobile device and shows up as a shortcut on a listener's phone, allowing them to easily and immediately tune into the station's web stream.

Sure, folks can visit your station website whenever they like and play it on their phone, but there is nothing like living on their home screen and becoming an icon in their daily lives. Getting folks to download your app is like giving them a key to the 24-hour party. They can click in whenever they like, no delay, no searching around.

To create an app, you can either build one yourself or pay someone to build one for you.

REMEMBER

Building your own app allows you to be in control of all aspects of your application, including what is advertised and what sort of data is being collected on your listeners. It also allows you to make updates as necessary and create a customized user experience.

You don't need to pay a lot of money to create your own app. In fact, at WXOX, our volunteers Mathias Kolehmainen and "Wild" Bill Haynes had no experience building apps, but they both have great technological minds and were willing to take the plunge. The result has worked out very well! The WXOX apps are very straightforward without any bells or whistles, but they work! I think this is what users like about them — the simplicity — no adverts, no distractions.

TIP

If you're interested in trying this out for yourself, the Apple developer site offers free instructions for building IOS apps using Swift. To get started, visit `https://developer.apple.com/learn`. In addition, *Android Application Development All-in-One For Dummies* by Barry Burd and John Paul Mueller (John Wiley & Sons, Inc., 2020) provides step-by-step instructions for building your own Android applications.

If you don't have the time or know-how to build your own app, a number of online providers will create apps specifically for your radio station. These include

- Aiir (`aiir.com/mobile-apps`)
- Listen2myradio (`listen2myapp.com`)
- Nobex (`nobexpartners.com`)

Some providers, such as Nobex, are free but insert advertising, and others will build custom apps with interactive elements for a fee of approximately $300 per year.

TIP

Put plenty of thought up front into which application builder you choose. Consider sound quality, visual appeal, ease of use, and user privacy. Look for examples of other stations that use the app. Try them out for a few days to see how they work before you commit. Does the music start and stop appropriately with the buttons (hopefully)? Does the music keep playing if you navigate away from the app (it should)? Does it ever act with a mind of its own, stopping automatically or starting up when it is in the background (it should not)? The last thing you want is an unruly app that frustrates your listeners. They will delete it and never come back. After you create your app and get it on people's phones, you will want to keep it active and responsive. Be sure to update as necessary.

Provide links to your apps on your website and social media, and in all of your digital correspondence. You may want to put links to your apps in your email signature. Make sure that you have plenty of on-air announcements letting listeners know how they can download your station app. This is a great way to move your listeners away from the other commercial listening platforms into the local fold.

A number of LPFM stations now have their own apps; you can see a few examples in Figure 5-6.

FIGURE 5-6: Many LPFM radio stations have their own apps, a number of which are pictured here.

WARNING

Developing the free WXOX app was one of the best decisions we ever made for our station, but we did it in the worst possible way. Our stream host was offering very inexpensive custom applications as an add-on to their streaming service. It was a one-time fee of $60 which was wonderful — so, why not? Well, we learned why not a couple of years later when they suddenly discontinued the service but left our nonfunctional app online. We tried so many ways to contact the developer to have it either updated or taken offline but to no avail. We subsequently developed another app that worked perfectly, but for years it sat in the Apple Store next to its sad, lifeless sibling. I assure you, it caused plenty of confusion. The moral of this story is: Always make sure that you are in control of your own station app.

Teaching Skills to Smart Speakers: "Alexa, Play WXOX"

Alexa is a voice-controlled virtual assistant operated by Amazon. Individuals with Alexa-connected devices can speak commands that activate tasks programmed into her operating system and she responds accordingly. These commands are called *skills*. You can develop a skill for Alexa that teaches her to play your station's web stream on command.

Although it used to take developers months to get Alexa Skills designed and approved, an easy on-boarding process is now in place for radio stations that teaches Alexa how to play their streams directly. Stations are added through a simple intake process that provides Alexa with the streaming URLs and necessary metadata. When authenticated, the station audio plays when users say "Alexa, play <Station Name>." Alexa is able to play stations by call letters, alternative names, and frequency.

TIP

To get started, send an email to `rsk-onboarding@amazon.com`. Almost before you know it, Alexa will be playing your station!

3

Staying in Orbit: Building a Station That Endures

IN THIS PART . . .

Bring your studio to life by finding a great location, installing the right gear, and getting your community on board.

Curate an amazing audio collection for your station in both digital and physical formats.

Safeguard your station's future by securing your finances through fundraising, membership, partnerships, and sponsorships.

Budget successfully by factoring in all of your outgoings, from paying the bills and buying technology to music licensing and paying your operational staff.

IN THIS CHAPTER

- **Finding a great studio location**
- **Selecting and installing the right gear**
- **Inviting the community inside while keeping everything safe**

Chapter 6

Where the Magic Happens: Building a Studio with a Spark!

Building your studio is super exciting — you should have a lot of fun as you begin putting together the pieces with which you'll create your living audio laboratory (in other words, your radio station!). In this chapter you discover what to look for in a physical location, how to select the right gear, and how to put it all together to create a bumpin' studio atmosphere. Finally, I discuss the importance of making your station headquarters accessible to the community and safe for volunteers, while protecting your hard-earned equipment.

TIP

Take your time building your studio; leave it open to growth and development. Keep in mind that technology may change and you want your studio to endure. Spend money on the environment and the lighting. Bring in flowers from time to time, or decorations for the holidays. It is important so set the right mood and create the right vibrations!

Finding a Studio Location

Studio rental can be one of the greatest expenses for a LPFM radio station. It is not unheard of, however, to find a landlord that is willing to donate space to the community radio cause. You may find a family member that is willing to invest (thanks Nancy!). You may find someone or some business that would be thrilled to have you announcing a few times a day that you are transmitting from their location. In fact, a business would be foolish to turn down this rare and powerful opportunity!

Always be careful when partnering with another organization. Make sure that your values align and that they will not attempt to control the content of your programming. Make sure there aren't any character issues with the landlord that could cause problems down the road. Even if the space is donated for free, get a lease drawn up where you pay at least $1 a year as assurance that you will not get evicted without warning or be hit with any surprise maintenance or utility fees.

Because you are a radio station, think about noise and your neighbors. ART FM began in a donated studio space in a residential building. The landlords were very generous but out of respect for the other tenants we were always telling the DJs to "sssssshhh" and "turn it down." Forcing a radio station to act like a library is torture for everyone involved. Ultimately, a live band after hours brought it all to a sudden end. Luckily, fate introduced us to a new free-standing building on the very next day. Now we make all the noise we want! (Check out the nearby sidebar for more on this story.)

The more space you have the better, but not a lot of room is required to build a radio station. In fact, one of the largest noncom stations in Louisville got started in a broom closet at the public library, running microphones out of the door to live musicians playing out in the hall. The point is to make it work; get it started in your basement if you have to! Eventually new opportunities will arise.

Above all else, you want your studio to be safe for your volunteers who will be coming and going 24 hours a day. Be careful of stairs or other impediments that may make it difficult for those with mobility issues to participate. Make sure that the area is well lit and far from any known trouble zones. You want convenient parking for your hosts and guests. Your personal travel distance and traffic issues should be taken into consideration, too, as there will be times you need to get to the studio in a flash.

Your studio equipment will be the same whether or not the antenna is on-site. If it is, you will have the transmitter and EAS equipment in house. This makes troubleshooting easier as everything is located in one place. If your antenna tower is in a different location than your studio, your EAS equipment and transmitter will be housed at the remote location. For more on this transmission chain see Chapter 4.

When seeking a studio site, be aware of FCC localism as discussed in Chapter 2.

FROM FOLDING TABLES TO A "SPECTACULAR RADIO STUDIO"

ART FM got our start on folding plastic tables in pop-up locations around the city. We would lug our equipment around from night clubs to museums, to community picnics, setting up whenever there was something interesting going on. Couch-surfing radio style! Our nomad radio station finally found a home in April 2013 at Piazza de Felice — an unoccupied commercial property in the Nulu artist district — which was generously donated by an ART FM board member. It was so good to have a space of our own! It was here that the station really got going but as a music-making factory it was always a struggle not to disturb the residential neighbors upstairs.

The very day it was determined we needed a new home, we found a free-standing building in downtown Louisville where we could make as much noise as we wanted! Through a contest conducted with the help of the Louisville Visual Art (LVA), we selected Wilfred Sieg of The Art Cartel (`theartcartel.org`) to paint a mural on the side of our building. Like our signal, it's a colorful, musical addition to our neighborhood and now you just can't miss us!

A few years after moving into the new building, our DJs raised the funds for a new console during "Give-For-Good Louisville" a city-wide day of nonprofit giving that was held before the COVID-19 pandemic. The board of directors then agreed to match their work with additional equipment and a complete studio remodel. The plans were drawn up in late 2019 but before we were able to get to work, the pandemic arrived in the United States.

As soon as we heard about stations on the West Coast shutting down due to the pandemic, we got busy teaching our DJs how to live remote. By the time there was a health emergency in Louisville, all of our DJs were broadcasting live from the safety of home (find out more in Chapter 15). During the time the WXOX studio was shut down in early 2020 we had an opportunity to do a whammy of a facility remodel.

As we had moved into our building just before our FM launch in 2016, we were very rushed and did not even have an opportunity to put a new coat of paint on the walls.

(continued)

(continued)

Prior to its use as a radio station, the building had served as a theater prop company, F.O.P. headquarters, and most recently, a hair salon. There was literally hair dye all over the place. In some areas, it looked like a crime scene. Therefore, getting fresh paint and new carpet was a top priority in this studio remodel. We also replaced the overhead fluorescent lighting with dimmable incandescent track lighting. With these improvements, the environment was immediately transformed!

We busted out a window between the studio and an adjacent performance room. The original plan called for a glass sound-barrier to be placed in the opening. However, in response to the pandemic, we left it open to provide better air circulation within the studio area.

From the beginning, we planned to move the old equipment into another room at the back of the building and construct an off-air production facility — Studio B. However, the persistence of COVID-19 upgraded the importance of having a second on-air studio. Making Studio B live was a simple process of running a Cat5 Ethernet Cable through the ceiling into Studio A and plugging it into the main console. The ability to rotate between studios allowed us to stagger shows between the two studios during the pandemic and reopen the station to in-house broadcasters.

The ART FM studio remodel is an ongoing process; the station is constantly evolving. In 2023 WXOX was named a "Spectacular Radio Studio" by *Radio World Magazine*. This is a great honor, but no doubt we agree! Our station is featured among 15 other fabulous international radio stations. There are some really beautiful studios in there but none, in our opinion, compare to the warmth, comfort, and creativity found within the walls of ART FM!

Considering Interior Spaces

The radio station is your clubhouse, so you want it to feel comfortable and inviting. Guests should feel relaxed when they enter and they should also feel impressed. That way, they will want to show off that they are there. It's always good to have your logo on the studio wall somewhere as a "step and repeat" for guests to get their photos taken in front of for social media purposes.

Of course, the amount of flexibility you have with your setup will depend on the amount of space you have to work with. If you have your choice of rooms in which to put your main studio or control room you will want to think about sound. Are there environmental or mechanical noises that you will need to block out? How much sound conditioning will you need in the studio itself?

All of this will be up to your personal preference. There's nothing wrong with embracing environmental sounds or playing background music to block it out. (For more about how KAAY DJ Clyde Clifford used this to his advantage, see Chapter 11.)

Soundproofing

If you need to soundproof your space to keep the audio from leaking out and disturbing your neighbors, there are a variety of options. Most involve adding an additional layer of drywall around the room separated by an air pocket from the first layer that will absorb audio vibrations.

When selecting materials, investigate special products such as QuietRock (`www.quietrock.com`), Auralex (`auralex.com`), and GreenGlue (`greengluecompany.com`) designed to increase the soundproofing capabilities of your new construction.

Sound dampening

If you need to dampen the sound within your finished studio space and prevent noise from echoing around, you have a number of options. You can choose from a wide variety of decorative acoustic panels from commercial sources such as Acoustic America (`acousticsamerica.com`) or Acostimac (`acoustimac.com`). Alternatively, there are a number of YouTube tutorials for constructing your own acoustic paneling. You can also simply attach egg-crate foam to the walls or hang thick curtains.

Adding wooden cubbies such as music shelves can also help cut down on the sound reflection, especially if they contain vinyl records with absorbent cardboard covers.

Installing Studio Furniture

Each radio studio will have its own needs and requires a custom layout. As crazy as it sounds, I absolutely recommend starting with plastic fold-out tables. They cost around $40 apiece at your local hardware store. Four- and 8-foot tables come as standard (as do tablecloths of the same size). Grab some folding chairs and *voila*! ART FM worked like this for a while in the beginning and we found it helped to move things around and try different setups before locking it down. Meanwhile, you suffer no financial loss on the folding tables and chairs after the upgrade, as they will continue to come in handy for a variety of station events. Perfect!

TIP

As far as the actual setup goes, I like a U-shape studio set up where the main audio console is straight ahead. The turntables and equipment rack are off to one side of the U and open desk space for interviews is on the other. In Figure 6-1 you can see the configuration we used in the WXOX remodel. Eventually, the DJs asked for more counter space to lay out records, so we built a cart on wheels that could be moved around the studio to the most convenient position for the on-air DJ.

To be sure, a DJ should always be encouraged to dance! As such, I am a fan of a standing studio but do think it should accommodate folks who wish to sit down as well. Therefore, counter-height tables make the perfect middle ground. Countertops that are 36" tall allow your DJs to easily stand up while on air, move freely about the area, and still control all the equipment comfortably. Get countertop-height chairs, so that program hosts can sit down and manage the audio controls very efficiently as well.

TIP

Chairs are one of the most well-used pieces of studio equipment and you should expect that they will get treated rough. Metal construction lasts longer than wood. Make sure the studio chairs do not have arms that could hit the turntables or otherwise get in the way. If you're on a tight budget, start with stools found at restaurant supply warehouses. These are nice and durable. After you have a bit more to spend, you can invest in better-quality counter-height studio chairs. At WXOX, we have found the 9 to 5 Seating Logic 10" stool to be a strong and good-looking choice. It comes in a variety of custom fabrics. We currently have four of these in our studio. Two have been around since the very beginning.

TIP

When you are ready to settle into your studio design, I encourage a custom build from a local carpenter. You should get some bids and put this into your fundraising budget (see Chapters 8 and 9 for more on budgets). Racks for holding equipment, channels for running wires, openings for connections and panels for quick tech access will make your radio life much easier in the long run. The wrong setup can be a big mess of a lot of wires further down the road. If you have seen Terry Gilliam's film *Brazil* then you know what I am talking about! The right carpenter, however, can help you design custom solutions that will make your studio shine and hum. WXOX is lucky that Sean Selby (co-founder and chief operator) moonlights as a master carpenter. His skills have blessed us with a very beautiful, highly functional station. We have fake walls, secret doors, and hidden chambers all built into our control room furniture. Not only is the installation perfect for our radio needs but the custom woodwork has a warm, delightful feel.

It's a good idea to visit as many radio stations as possible when designing your home base. College and community stations are usually very approachable but emailing station management to make arrangements is preferrable to showing up unexpectedly.

FIGURE 6-1: The WXOX studio remodeled by ART FM chief operator and master carpenter Sean Selby. Designated as a "Spectacular Radio Studio" by *Radio World Magazine* in 2023.

© Nancy Scott Photography.

TIP

Visiting distant stations can cost time and money that you may need for other purposes while you are getting started. Luckily, Jennifer Waits, co-founder of *Radio Survivor*, has done the research for you! Her tours behind the scenes of over 150 college and community radio stations have been posted to the Radio Survivor website (radiosurvivor.com/radio-station-tours-in-numerical-order). Here you can check out a variety of unique station setups that were built on a budget with longevity in mind. A number of Jennifer's photographs are included in this book.

Selecting Studio Equipment

REMEMBER

Audio equipment will be your station's biggest investment. Think long term and look for gear that is durable and easy to use. Focus on the essential pieces and move forward slowly from there.

For the best selection, look for retailers that specialize in broadcast gear including Broadcast Supply Warehouse (bswusa.com), B&H Photo (bhphotovideo.com), Sweetwater (sweetwater.com), and Broadcasters General Store (bgs.cc). They have equipment specialists able to answer your specific questions regarding installation and on-air use.

Audio console

REMEMBER

There are a few essential pieces of equipment at any radio station. First and foremost is the *audio console*, also known as the *mixer* or *soundboard*, or simply *the board*. It controls all the components and the audio levels they send out to the listener. The audio console is the brains of your station it is the place where everything comes together — truly, where the magic happens!

A number of companies manufacture broadcast consoles, including Wheatstone (`wheatstone.com`), Arakkis Systems (`arrakis-systems.com`), and AEQ Broadcast (`aeqbroadcast.com`). If you want something handmade, engineer Al Davis at the Kaatskit Radio Company will build a modular console specifically to meet your studio needs (find out more at `kaatskit.com`).

TECHNICAL STUFF

Radio consoles have a few features that are different from other audio mixers. They have built in mic channels that will silence the in-studio speakers when they are activated, to prevent feedback. Likewise, they will have a "cue" button for each input that allows you to preview upcoming music while the original source continues going out to the listeners uninterrupted.

Consoles are offered in two main varieties, analog and digital:

- Analog boards have been standard since the dawn of radio. They use solid-state electronics and every variable has a physical control. Some people prefer the sound of analog boards to digital ones.
- Digital boards allow for many more inputs and customizations than analog boards. However, it won't take you long in the community radio biz to realize that infinite customization is not what you are after. With a host of individuals with limited technical understanding you will find that the more options you have, the more opportunities there are to mess things up!

TIP

Strong, simple, straightforward should be your mantra when searching for studio equipment.

After some debate, ART FM went with an analog sound board. We began with the AudioArts Air 1 by Wheatstone for the WXOX studio, which, in my opinion, is the perfect starter gear for a LPFM station. It is an affordable, friendly little board with simple, straightforward controls. It is easy to use and it sounds great, too! The only drawback is that it has limited channels. There are only two designated mic channels and six audio inputs. We survived like this for years and only recently upgraded to the AudioArts Air 5. It functions exactly like the Air 1, which made for an easy transition, but now we enjoy four mic channels and 12 audio inputs.

REMEMBER

If you plan on webcasting, life will be easier if you acquire a soundboard with a USB output with which you can feed the audio directly to your streaming computer. Without a USB out, you'll need to provide the streaming computer with its own analog audio connection. There's lots more about setting up a webcast in Chapter 5.

Another option to think about when selecting your console is the button configuration. I prefer on/off buttons below each input fader. Sometimes these are located above the faders or they may not even be there at all — because sliding the fader all the way down effectively turns it off without the need for a switch. To me, however, there is nothing like the feeling of punching that input and dropping a killer song!

Microphones

If the microphones (or *mics*) in your community radio station aren't getting a lot of use, you are doing something wrong. Like everything in your studio, you will want them to be durable and user-friendly. Since all mics have a particular sound, I recommend that you use the same microphones throughout your studio build.

Microphones are divided into two main varieties, dynamic and condenser:

- Condenser mics are more sensitive to musical instruments but they are also more delicate and require external *phantom power* (DC power that activates the electronic circuitry within the microphones).
- Dynamic mics may not be as sound sensitive but they are the broadcast standard because of their durability and ease of use.

TECHNICAL STUFF

Consider the directional pattern of your microphone. Studios use omni-directional, shotgun, and cardioid mics. Omni-directional mics pick up sound from all directions equally, so you will get a lot of room noise. On the opposite end, shotgun mics isolate the area right in front of the microphone. The cardioid mic is somewhere in between; it has a heart-shaped pattern which hones in on the area in front of the mic but also picks up some audio off to the sides. Cardioid mics are good for community stations as they provide a focus on the speaker's voice but are a little more forgiving if the speaker moves off to the side.

TIP

At WXOX we went for Sennheiser MD 421-IIs (`en-us.sennheiser.com`) which are dynamic mics with a cardioid pattern. They were made for recording instruments but are versatile enough to handle voices smoothly. They are on the more expensive end, but we have been happy with their strong, robust sound. They have been fairly good for the community setting, although there is a release button on their casing that makes it too easy to drop the mic out of its housing by mistake. We use a tiny zip tie to hold it in place.

Other mics used by community stations include ol' reliable, the Shure SM57 (shure.com), which is an inexpensive option known for its durability and is often used in live performances. The '57 is a good starter mic — as is its sister mic the SM58 which is a more vocal-focused microphone. If you are satisfied with the brand, you can step up to some more sophisticated choices such as the SM7B. The RODE Procaster (rode.com) was made specifically for podcast use and seems to be well liked by the stations that use them. The Electro-Voice RE20 was originally designed for voice-overs and is known for being able to handle loud vocals without clipping or distortion — which makes it a favorite for shock jocks and other sensationalist broadcasters who habitually scream into the mic.

Mic booms

The *mic boom* is an adjustable arm that holds the microphone. The boom allows DJs to position the microphone exactly where it should be so that they can speak into it comfortably. Booms come in a variety of sizes, shapes, and colors and range in price. This is an area I would not recommend skimping in. A cheap mic boom will squeak and moan when the talent adjusts it on air and this can be quite embarrassing.

In the WXOX studio upgrade, we went with O.C. White (ocwhite.com) booms. They are not too expensive and have been working well with no squeaking yet. Yellowtech (yellowtech.com) calls their line of Mika booms "the world's most beautiful mounting system" and from what I can tell, they are not lying. These are recognizable in any studio that uses them and really give it that "wow!" factor. I have not been fortunate enough to try these out yet as they are quite expensive.

On-air light

On-air lights may seem like a minor detail but they are actually very useful for letting everyone know when the microphone is live. The broadcast console will have a tally that triggers the on-air light when the microphones are turned on. At this point, everyone knows it is time to choose between shutting up or shipping out.

You can buy regular on-air lights through most broadcast suppliers such as BSW (bswusa.com) or Sweetwater (sweetwater.com). Custom on-air lights made by Sandies are available through the same broadcast equipment dealers. Get some with your call letters on them! They aren't too much more expensive than the regular build and I promise they will bring a smile to your face every time they light up!

Headphones

REMEMBER

Every time the mic goes on, so do the headphones. If you see a broadcaster talking on air without their headphones, you know they are a novice. It is important for a person to know how they are sounding on air.

During the COVID-19 pandemic, we required DJs to bring in their own headphones and mic covers to the studio. This seems to be a good policy and — so far — we're sticking with it. We do have a few extras of each stashed away for guest use — or if the DJ forgets their own — but we heavily stress the importance of everyone having their own personal gear.

WARNING

It is imperative that the studio headphones do not leak sound that will cause feedback in your mic. Avoid open-back headphones. Most closed-back headphones should work just fine and they do not have to be expensive. You do not need to go all the way for noise-canceling headphones as it is important to be aware of your surroundings.

There are many headphone choices from Sony, JBL, Phillips, Beats, and more. Audio-Technica makes good, closed-back headphones that we use in the ART FM studio. The ATH-M40x can be found in five-packs bringing them to less than $100 each. One great thing about these headphones is that they come with interchangeable cords — which is helpful because the cord is always the first thing to break! These headphones come with a straight and a curly cord. I prefer the straight cord in studio as the curly cords get tangled up with one another very easily. You may want to add hooks to the studio wall or under the console desk where the headphones can be hung when not in use.

TIP

Nearly all consumer-grade headphones have $\frac{1}{8}$-inch plugs, while nearly all broadcast equipment has $\frac{1}{4}$-inch jacks. Therefore, your studio will need adapters — a *lot* of adapters. If anything walks off from home base, it will be these; it's easy for the DJs to forget they have them on their headphones and end up with an unnecessary collection at home. Just ask and they will bring them back.

Headphone amp

Your soundboard will have one headphone jack but obviously, you need more than one pair of headphones. Therefore, you need to connect an external *headphone amplifier*, which allows you to get eight or more headphone channels through a single external amplifier. Each channel on the amp has its own volume controls so the DJ can accommodate guests of various audio preferences.

You can even take this one step further by introducing individual *talent boxes* at each seating area, providing the guest with a personal headphone volume dial and possible "cough button" to mute their microphone when they need to cough or clear their throat.

At ART FM we use the ART HeadAmp6 Pro in our main studio and have also used the less expensive Behringer HA6000 for years. Mackie makes the MH-800 which will likely be our next choice in a headphone amplifier.

Studio monitors

In the world of pro-audio, speakers are called *monitors*. The control room monitors will not affect the sound of your broadcast but may it affect the mood of the DJ. If the radio host is really feeling the music they are playing, they will express it on air and the better the show will sound to those tuned in at home. You want your DJs to enjoy playing music in-studio so do what you can to invest in a good-sounding sound system.

REMEMBER

One of the features that makes the radio console distinct from other audio mixers is the ability to connect two sets of speakers — one to your in-studio monitors, and one that can go to another listening room. When you push the cue button on your soundboard, it activates the control-room monitors to play the music you are previewing, but it will allow the speakers in the listening area to continue playing what is going out over the air. This is really convenient for the DJ, who can hear what they are cueing up in the foreground while listening in the background to make sure their broadcast is sounding good. Therefore, to use your soundboard to its full potential, you will actually want two sets of monitors.

TIP

Listen to a variety of monitors before purchasing. In addition to local stereo dealers, Guitar Center (`guitarcenter.com`) stores have demonstration areas in most midsize cities. Likewise, national trade shows such as NFCB (National Federation of Community Broadcasters) and NAAM (National Association of Music Merchants) offer good opportunities to check out the latest gear.

At ART FM we chose the M-Audio Bx5 D3 (`m-audio.com`) monitors for our control room and the M-Audio Bx8 D2s as the secondary set for our lobby. These are nice-sounding, robust speakers that carry the bass well and never allow the treble to get too screechy.

Turntables

As soon as we decided to build a radio station, the first order of business was to purchase a pair of turntables. Vinyl DJs need two turntables to move back and forth between them; while one record is playing, the next song is getting cued up on the other turntable.

At ART FM we had no doubt that we needed Technics SL-1200s to set things off on the right foot. Produced by the Japanese electronics company Panasonic, these are the industry standard, or at least they were until they went out of production in 2010. Luckily, these turntables are tanks and there are many pre-loved machines still in circulation. The 1200s, as they are affectionately known, have *direct-drive motors*. In these machines, the motor shaft is directly connected to the base of the platter that holds the record, making it turn. This is superior to turntables driven by belts which can get stretched out, affecting the quality of play.

In 2016, Panasonic returned to their senses and began reissuing the Technics 1200s (`us.technics.com`) in limited quantities. These new releases are still direct drive and based on the original technology. They are shiny and colorful but they cost twice as much as the used decks, clocking in at $1,000 each. So it's still worth hunting down high-quality used items or seeking alternatives. A few options mimic the Technics design and these include the Pioneer PLX turntables, which don't offer much of a price break from the 1200s but they do have advanced pitch control features. And the Audio-Technica AT-LP line of record players is a much more affordable turntable option.

TIP

Most radio stations have two turntables. When selecting right pair for your studio be sure they have the features that DJs need:

- **Motor:** Direct Drive for scratching, quick response, and longevity
- **Speed:** 45 and $33\frac{1}{3}$
- **Platter:** Manual start and stop
- **Tone arm:** Movement independent of platter
- **Pitch control:** Ability to match speed

FINDING THE FABLED 1200S

After production ceased, the only way to get your hands on a pair of 1200s was to purchase them used. Sure they're back in production now, but new ones cost so much more than used examples. The great news is that even older, beat-up looking decks tend to play nicely when they are given a little love. Don't worry if yours is missing things such as the stylus, head-shell, cover, or 45 adapter. All of these are readily available from sources such as `1200s.com` that specialize in replacement parts for Technics turntables.

(continued)

(continued)

If you find a used player, it is a good idea to take it to a local electronics shop and ask them to give it a tuneup. Depending on its condition, you should be able to spend around $100 to get this done — of course, necessary repairs and replacement parts may end up costing more.

1200s show up regularly in used equipment shops and if you keep your eye out, you should be able to find them for less than $500 apiece. Check with local electronics businesses first but national outlets such as Music Go Round (`musicgoround.com`) and Guitar Center (`guitarcenter.com`) may have the lead on used players in other parts of the country. Reverb (`reverb.com`) is an online marketplace where individuals list new and used music gear for sale. Technics can be found on these websites for good prices. I'd recommend only purchasing unseen equipment if you can return it if it arrives damaged or defective — and watch out for return shipping costs.

When we were purchasing our turntables for ART FM, we saw some pretty messed up gear. Remember, these machines are often used by club DJs who drag them around to different parties night after night. It is common for them to have bumps and bruises. That can be okay so long as the mechanics are in working order. One of the first tables we ordered used for ART FM arrived coated in a fragrant green crumbly dust that had gotten inside to the gears and moving parts — that must have been quite a party! Luckily, we were able to get a full refund and then try another.

CD players

REMEMBER

Keep it simple! I can't reiterate this enough when it comes to studio equipment and I keep learning this lesson the hard way when it comes to CD players. Leave the fancy features aside — skip the Bluetooth and MP3 capacities. The only controls you need are a simple way to scroll through tracks, and to start and stop the player at the appropriate moment.

The display should be as straightforward as possible. It will need to have an easy way — one button — to find the count down, the "time remaining" on any given song. It is always important for DJs to know how much time they have left. At ART FM, our Tascam CD200s have been golden. During the studio remodel, we "upgraded" to more expensive Denon DN-700CB players but ended up coming back to Tascam for simplicity and ease of use.

REMEMBER

Radio stations usually choose CD and cassette players that are rack mountable and attach them to an equipment rack built into the studio furniture.

Cassette players

You'll see a few varieties of tape players in community radio stations. Most of the time, these are cassette players but you still see cart machines and reel-to-reel players from time to time.

Cassette players should also be bare-bones. Open/close; play/stop; fast-forward/rewind; pause. That's all you need. The fast-forward and rewind functions should have a "seek" feature that can stop in between songs. It is nearly impossible to cue tapes without this. Avoid the automatic flip feature where the player plays the opposite side of the cassette to the side facing forward. This can make things very confusing! Double cassette decks can be okay and have the bonus dubbing capacities, but two individual machines will allow for smoother audio mixing through the console.

Presently in the WXOX studio, we have a single Denon DN-790, which has a high-fidelity sound and low motor noise. We did not originally have a cassette player in studio (I mean, who plays cassettes these days?) but the DJs clamored for it and we obliged. It does seem to get use — mainly by the younger folks who never *had* to use cassettes in the first place.

Computers

Mac or PC? It is a decades-old debate but it truly does not matter which platform you use for your radio station so long as it works for your applications. This is totally up to your technology preference. In fact, it is not necessary at all for an FM station to have a computer in-studio, and stations functioned without one for years. However, they can make life much easier.

At WXOX, we have three Mac-Mini computers in our main control room. Each of them has a different function: webcaster, archiver, and player. Here's a look at each function in turn.

Webcaster

If you wish to stream your audio online you will need a computer that receives a feed from the soundboard (usually by USB input) and uses audio encoding software to send it to your stream host (see more about this process in Chapter 5). We keep this computer focused on its one job and it is 100 percent off-limits to the volunteer staff unless the system crashes and it needs to be rebooted.

Archiver

If you want high-quality recordings of your broadcast, you need a computer that takes the audio directly from your soundboard. The archiving computer and the streaming computer can be one and the same, but at ART FM, we prefer to use two separate computers. This way, the DJs can record their programs without messing with our stream computer. Audio Hijack (`rogueamoeba.com`) is installed on both computers acting as our encoding and recording software. Other recording options include Audacity (`audacityteam.org`), Garage Band (`apple.com`), SoundForge (`magix.com`), and Adobe Audition (`adobe.com`).

Player

Most stations have a computer that feeds audio to their soundboard. At ART FM we call this computer DJ H.A.L. because we use it for our automated playlist using the program Mega Seg. You can find out more about the options for managing your automated programming (and DJ H.A.L.) in Chapter 7.

Some stations allow their DJs to play music from this computer but at ART FM we need to be sure the music in our auto-mix is 100 percent clean and the remote schedule is not interrupted. Therefore, keep this system off-limits from the station volunteers. The DJs know they can use this computer to access bookmarked pages that include essential items like station announcements but they should not touch this computer otherwise! If a DJ needs to use a computer, they must use the archival computer instead.

TIP

In Chapter 9, I discuss the importance of a system that will allow you to access and control your station computers remotely. ART FM has a remote log-in system installed on our main computers, and this has saved a lot of late-night trips to the station.

Computer cables

A station should supply two $\frac{1}{8}$-inch stereo cables to connect computers, smartphones, mixers, and tablets to the soundboard. These computer or *comp cables* plug into the $\frac{1}{8}$-inch headphone jacks on the electronic device in order to bring its audio on air. Most DJs will want to connect their own electronics at some point, whether to play music or to conduct an interview. These comp cables make it possible. They do go bad after a while and may get a glitchy or one-sided audio signal. So be sure to have plenty of extras on hand.

You may want to run the comp cables through a balance box before they reach the soundboard. This is not necessary but the balance box converts the unbalanced signal coming from the electronic device into a balanced signal for the soundboard to receive to create a better sound by canceling out unnecessary noise. At WXOX, we have had good luck with the Henry Engineering Twin Match. One "twin" box can accommodate two comp cables, each running to their own channel on the soundboard.

Accessories and adapters

All the big tech stuff featured so far in this chapter will do you no good unless you have a way to connect it all together! It doesn't seem exciting to think about cables and adapters but it sure is important.

When you order your equipment, speak with your sales rep to find out what kind of connections you will need. Some equipment will come with more installation gear than others. Some will require parts that you need to prepare and install yourself. Make sure that you can handle the expectations. If you are not an audiophile, chances are you can find someone in your community who enjoys assembling stereo equipment and who is willing to help you out. It's not often a person gets to help build a radio station!

All broadcast studios need to consider surge protectors, extension, cords, and power strips that protect and connect their equipment. Will you have a backup power supply? While a generator may not be in your budget yet, you may want an Uninterruptible Power Supply (UPC) to keep your electronics online through short power surges or brief outages.

Distribution amplifiers are devices that accept one or more audio inputs and provide several outputs. You can use the outputs to feed audio to things such as studio monitors, the web stream encoder, the archive computer, and so on. Many stations use distribution amps in order to avoid using "Y" cables (these cables split the feed into two identical signals but can degrade the audio quality).

Many radio stations have a dedicated *engineering office* for repairing equipment, soldering, and housing equipment that doesn't need to be in the studio. Having a room with a workbench may seem a luxury, but is nearly essential to the smooth and continuous operation of a station.

Providing Community Access and Security

After you fill your studio with all this hard-earned, well-considered equipment, you will want to be very careful about who has access to it. Your DJs must be well vetted and well trained. They must respect the space and care for the equipment — so they must understand how hard it was to get it! Any security systems you employ at the station are to protect your volunteers first and foremost. The protection of the equipment is secondary but also important to maintaining your operations. There is a plethora of DIY security systems available should it feel necessary. Just remember that these can sometimes use lot of bandwidth that you may not have to spare. Too much activity on your network can cause your web stream to crash or become glitchy.

You will need to decide how your volunteer DJs get access to the studio and how you monitor anyone else who comes and goes. WAYO-LP, founded by graduates of the Rochester Institute of Technology, developed their own key fob system that tells them exactly who is coming and going and when. Most security companies offer custom key fob systems that do the same thing, but these can be expensive. You can also use a simple keypad system. Schlage (`schlage.com`) offers a variety of keyless locks that can be installed yourself. Be sure to change the code every time there is a personnel shift. It is important that at any given time only the current staff members have access.

WXOX EQUIPMENT: STUDIO A

Below you will find a list of equipment from Studio A, the main on-air studio at WXOX. This information is provided as an example of the type of gear you will need in your own studio. Each station will make their own custom choices but we have been happy with all of this equipment for its simplicity and durability — key factors for the community radio setting. A more detailed version of this list available in the online resources of this book.

- Custom furniture and shelving by Sean Selby Carpentry
- Console: Audio Arts Air 5 by Wheatstone
- Monitors: M-Audio Bx5 D3
- Microphones: Sennheiser 421-II
- Record players: Technics 1200 MkII and Technics 1200 M3d
- Turntable pre-amps: Radial J33

- Cassette player: Denon DN-790 R
- CD players: 2 x Tascam CD200s
- Balance boxes: Henry Engineering Twin Match
- Comp cables: 2 x for connecting DJ computers
- Headphone amp: ART HeadAmpPro6
- Headphones: AudioTechnica ATH-M40x
- On-air light: Sandies custom "ART FM"
- Computers: 3 x Mac Minis dedicated as follows:
  - Automation computer running MegaSeg by Fidelity Media
  - Streaming computer: Audio Hijack by Rogue Amoeba
  - Archiving computer: Audio Hijack by Rogue Amoeba
- Back-op power supply: APC UPC 1500VA
- Site to transmitter link: GatesAir Intraplex IP Link 100 to BW 300V3 transmitter

IN THIS CHAPTER

» Building a music library

» Organizing digital files

» Curating your in-house archive

Chapter 7

Cranking It Up: Developing Your Music Collection

When developing your music collection, the first question to ask is what sort of music your station will collect. Straightforward though it seems, it requires two responses. First, will you collect all genres of music or will you have a specific focus? At WXOX, we cast as wide a net as possible to accommodate the diverse interests of our DJs, but your station may want to specialize in a particular genre such as jazz, classical, or punk rock.

Next, you will need to ask which format you will collect this music on. Some stations simply keep a digital archive on their station computer while others house records, CDs, and tapes. Invariably, the answer to this question will depend on how much space you have to give it.

In this chapter, you will discover how to develop a living music collection — physical and digital — and have it grow with your station as it provides your DJs with great listening options while increasing the assets of your station.

Building a Physical Library

Most DJs bring their own music to the station, so it is not necessary for your station to have an on-site collection. However, having music in-house can be an outstanding resource for your DJs and provide them with access to music that they — and your listeners — may not otherwise have the opportunity to hear. You can curate your collection to create a particular sound for your station or you can take a broad approach that provides your DJs with access to a world of musical styles. In either case, an in-house library introduces your staff to new artists and more obscure musical releases. It is also essential for DJs who show up at the studio empty-handed but ready to make radio happen!

TIP

In terms of developing a valuable collection for your station, you need to think about what musicians may visit your studio. This is not crazy talk. Musicians crave airplay, and when you get going, they will seek you out. When they are there, it is easy to get them to sign your records and increase the value of your albums exponentially in a single instant. So, which musicians live in your community? Recently Paul Simon and Edie Brickell made an appearance at Wimberly Valley Radio (`https://www.kwvh.org`), a LPFM station near their home in Texas. Which musicians do your DJs have personal connections with? WXOX has a connection with punk rocker Lydia Lunch, and she stops by the station when she is in town and has even hosted a few guest radio shows. So, of course, we have Lydia Lunch records in house — available for her to draw on whenever she's feeling creative!

Station records will get played, hopefully, so you cannot expect them to be in pristine condition. Instead, you will have to approach collecting from the completely opposite direction by encouraging signatures and customizations from visiting celebrities. These records can become valuable physical assets to your station. It may take decades for a station's music collection to attain any monetary value, or it can happen overnight with the visit of one important celebrity. Imagine if Madonna shows up at your radio station — and you don't have records for her to sign; you have just missed a great opportunity. Plus, how embarrassing!

RADIO CLASH: A LOST MUSIC COLLECTION

When Joe Strummer of the punk band The Clash landed in Nashville, Tennessee, the first thing he said was, "Take Me to WRVU." Or so legend has it. It was a cool spring day in 1984 when he showed up at the student radio station and began rummaging through the record collection, playing the piano, and conducting impromptu interviews with the star-struck student DJs. Before he left, he signed records held in the studio collection and drew funny images on the front of each one.

Years later when station alumni returned to WRVU, showing off the old stomping grounds to friends and family, one of the first things they would do would be to pull out The Clash records to reveal the carefree drawings Strummer had added to the front. The summer of 2010 was my turn to revisit my alma mater with my young family in tow. I couldn't wait to show my husband and son the station and specifically the Joe Strummer customized records. When we arrived, the studio was in the same place but downsized and completely remodeled to be utterly unrecognizable. "Where is the vinyl?" I asked the student DJing at the time. After a long pause, "In the dumpster" was her cold reply.

I was in a state of disbelief — not only over the loss of The Clash albums but also over the loss of an immense number of autographed records that had been incubating in that collection since its foundation in the early 1940s.

"The last of it is in the hall, waiting to go out," she continued. "They said we could take what we wanted." Sure enough down the hallway, in a space where we once caught an interview with Gwen Stefani, lay a cacophony of records strewn across the floor after being rummaged through by dozens of curious students. Some had little drawings and cartoons from the musicians on the cover.

I later confirmed with the station administrators that the albums had been tossed in the dumpster because they did not have room for them in the remodel and "the students don't listen to records." The stab went straight to the heart. Losing this record collection was the beginning of the end for this treasured radio station. A few months later, it was announced that the license would be sold to the local NPR affiliate. "Students don't listen to radio," they said when they blamed it on the finances.

The truth is that the vinyl treasures within this immense collection could have staved off whatever financial crisis the student radio station was facing. It alone could have prevented the sale of the frequency, but I am not entirely sure that is what the administration was after.

This experience did, however, make me recognize that the value of a radio station record collection goes beyond simply having great music in-house for the DJs to play. It's an investment that can provide a safety net for the future of the radio station itself, but only if it is done correctly. It's easy for collecting records to become a money pit, especially because the records will be used and mistreated by the station volunteers. You can't depend on the records being in mint condition at your station, but they are in prime position to be signed and customized by the musicians who visit your studio. You have to be ready.

Thinking about shelving (and space)

The single most important thing you can do for your music collection is provide proper shelving. Lots of cheap DIY solutions are available, but your collection will be heavy and you may soon find yourself with sagging shelves, lost pins, and so on.

TIP

I recommend finding a local carpenter who can build shelves specific to your studio needs. It should be a fairly straightforward process and hopefully not too expensive. Build what you can for now and plan to add more in the future.

REMEMBER

If you decide to keep an in-house music collection, you will need every wall for shelving. Every. Single. Wall. Maybe not now, but eventually. And then some. So think ahead. The greatest community radio stations, such as KBOO in Portland, Oregon, are labyrinths of shelving or "stacks" that run throughout the station grounds. It is a lot of fun to lose yourself there.

Magical things happen in the stacks. It was in the WEFT stacks in Champaign-Urbana that I learned that WXOX would get our LPFM license. Figure 7-1 shows how all of the walls are used, and how extras have been built to hold even more CDs. Above the studio window you will see a color-coded chart to musical genres (more on this in the next section). WEFT DJs have a number of folding ladders to access the higher shelves. I prefer classic library stools such as the Cramer Kik-Step that you can literally kick around the room when you are holding a stack full of records. The WEFT stacks double as an on-air performance studio.

FIGURE 7-1: The stacks/performance studio at WEFT Champaign-Urbana.

© Sharon M. Scott (Book Author).

Choosing an organizational method

Organizing the music collection requires some big thinking on the part of the station founders. Will the albums be alphabetized straight A to Z or will they be split into categories and then alphabetized within that category? Creating musical categories can be a very complicated process, especially with so many modern musicians blending musical genres.

Anticipating that this trend will continue, WXOX decided on a straight alphabetical approach based on artist names in our main stacks. We then identify certain genres with color-coded electrical tape on the spine of each record and sticky dots on the spine of the CDs. This system is flexible, and if the musician crosses several genres, you can add different colors of tape to represent each.

TIP

Create a chart that identifies the colors of each genre and hang it on the wall. Most community radio stations will have these if you look. Each one is unique; some get very complicated. At WXOX, we've tried to keep ours simple. Here is a list of the colors we use and what genre they correspond to:

- Red: Pop, Rock, Soul
- Orange: Jazz
- Dark Blue: The Blues
- Light Blue: International
- Yellow: Beachtime
- Brown: Folk/Country
- Green: HipHop
- White: Electronic
- Grey: Classical
- Purple: Odd
- Pink: Spoken Word
- Black: Local
- Gold: Station Family Bands

We determined that there were a few selections of music, such as our Holiday Music Collection and our Electronic Dance Collection, that needed their own specific shelving area. The key is to remain flexible, yet consistent, as you organize your collection, and the goal is to make it easy for the on-air DJ to use.

Collecting your materials and getting to work

You will need some materials to get your stations' collection organized:

- **Shelf divider cards.** Make your own shelf divider cards out of cardboard, though they will get a lot of use and plastic dividers will last much longer. Online, you can find custom-sized "shelf divider cards" for your records, CDs, and cassettes. They come with a variety of options. You can get them pre-printed with the letters of the alphabet or blank and use your own stick-on vinyl letters to customize. Even if you get the pre-printed A–Z cards, you will likely want a stack of blanks to designate any additional categories. Be mindful of which way the letters are printed, and make sure that they will read correctly when placed in your collection.
- **Vinyl letters.** You will want a package or two of vinyl letters that can adhere to the plastic dividers to designate particular alphabetical areas or musical categories.
- **Multicolored electrical tape.** Multicolored electrical tape will be useful in identifying musical genres. Try to get as many colors as you can at your local hardware store or by searching online.

  You will be surprised at the rainbow you will find! In addition to designating musical genres, the electrical tape is also useful for binding the spines of record sleeves that are split or worn.
- **Sharpies.** Sharpies should be in good supply at your radio station. You will use them all the time, including when writing on your album covers and the electrical tape used to create the color-key wall chart (outlined earlier in this chapter).
- **Covers and cases.** Depending on the condition of your music, you may need to order replacement record covers or CD cases. The collection needs to be in good working order. Nothing is more distracting to a DJ than trying to deal with a broken CD case.

Venturing into vinyl

DJs love vinyl. It's easy to rummage through quickly and it is easy to cue up (except when it is colored vinyl which makes it impossible to see the grooves). When building your record collection look for records that were recorded for that particular medium; you will find that they pop and fizzle in all the right places. I am not one to spend a lot of money on vinyl, but I do love to find treasures in the used bin.

REMEMBER

You likely will want different things for your radio station than most people want for their homes. Sound effects, historical recordings, and a sea of oddities can make for great radio when assembled in the right way. Recordings such as these can often be found at secondhand stores for just a few dollars. It is a great low risk way to get started!

Learning CD selectivity

CDs are great for DJs because they are easy to pop in and scroll through as you look for the perfect song. Even in this digital age, it can be nice to hold the physical disc in your hands, look at the cover art, and get a feel for what mysteries the music may hold.

TIP

If you have a listening area, DJs can arrive early for their show, grab a stack of CDs, and then take them into another room for previewing prior to going on air. In a nicely curated collection, they know that they can count on almost all of it to be good. It can be helpful to designate a few shelves in your station for local bands, new releases, or timely material that you wish to highlight on the broadcast. Figure 7-2 shows a nicely ordered, easy to navigate wall of CDs.

FIGURE 7-2: The CD collection at WXOX-LP Louisville.

© *Luis De Leon.*

If you are not careful, it is easy to get buried under the weight of too many CDs. You will immediately begin receiving CD donations from people in your community who have digitized their music collection. This can be a blessing and a curse. The important thing is to know when to say "No" or have an easy way to offload these CDs if they turn out not to be useful. Some stations give their DJs an opportunity to take these orphaned CDs home from staff meetings; other stations save them for a station rummage sale; and still others will donate them to local record stores. If you decide to give them to a secondhand store, I recommend that you donate them to a store that is out of the area so that the donor doesn't find them there and feel offended. Likewise, if you have marked these discs with your call letters, you will not want them in circulation locally as they will diminish your ability to recover stolen material.

When musicians begin hearing about your station, they will send you sample discs, and before long you will have so much new music that you won't know what to do with it all. The sad truth is that you won't be able to keep every donation. You must find a system for reviewing incoming music and determining whether it will be added to your rotation. This would fall under the music director's job description if you are lucky enough to have one (for more on various roles within the station, check out Chapter 12).

The music director will spend some time each week listening to the new CDs, maybe just skipping around and giving a minute or two to each of the tracks, to determine whether it is something that should be added to the station rotation. If not, the music director may choose to give away the promotional material, but they should not sell it.

The legality of selling promo copies has been hotly debated in court, and you should be aware that selling promos can get you blackballed from future promotional lists.

Tangling with tapes

Cassettes are making a comeback. They offer a rich analog sound and your DJs may enjoy having them to play on air. Personally, I find them high-maintenance and hard to cue up, but a few bands use the medium very creatively. As a result, WXOX has a small but significant cassette collection that focuses on music that has not been released anywhere else.

Programming albums

Many stations place a cover page on the front of an album or the interior of a CD cover to provide the on-air DJ with some information about its content. Most notably, "DO NOT PLAY" is written in big bold letters instead of song titles for pieces that violate Federal Communications Commission (FCC) law (see Chapter 10).

REMEMBER

This guide may also provide further info such as the length of the song, how it starts and stops, and what kind of impression it made on the person reviewing the album. This process is referred to as *programming* albums. Figure 7-3 provides a sample programming template. A customizable version of this template for records and CDs is available in this book's online resources.

****

Artist			***Album Title***	**Record Label**	**Year**
s1k1	***	Song Title	s/fade		6:52
s1k2	**	Song Title	f/false/fade		2:38
s1k3	****	Song Title	m/cool		11:55
s2k1	*	Song Title	s/segue		7:31
s2k2	* * * * * * * D O	N O T	P L A Y	* * * * * * * * *	
s2k3	****	Song Title	f/cold		7:17

This is one far-out record! Punk Rock but unlike anything I have heard before. Heavy on guitars and drums, yet mathy and melodic. Fairy-tale vocals with intense emotional dynamics on diverse instrumentation. The production on this album is first rate. Reminiscent of Slint, Rodan, and Rachel's with a touch of Massive Attack. Listen for subtle surprises.

DJ Jet Pilot 01/23

****

KEY:
Column 1:
S1k1 – side one, kut (or song) one
Stars – based on DJ appeal

Column 2: How the song starts/stops
s - slow
f - fast
m – medium
fade – fades out
cool – slow ending
cold – sudden ending
false - restarts after stopping
segue - blends into next song

Column 3:
Total runtime of song

FIGURE 7-3: A template for programming an album.

Often stations will allow DJs to check out records and CDs and take them home for programming. At the bottom of the programming sheet, the reviewer will share their impressions of the album and then leave a blank space for other DJs to hand-write their own impressions after they have played it on air. Over the years, these comments can provide fun commentary and valuable insight on the contents of the album.

Marking materials for safekeeping

To the horror of record collectors worldwide, scrawling station call letters on the front of record albums with a black sharpie is common business in the world of community radio. As you can see in Figure 7-4, some stations have stamps made specifically for the task. Not only that, most stations will also mark their call letters on the surface of the actual material as well — on the printed side of the CD or on the paper label affixed to the center of the record.

No, this not some sort of rock n' roll vandalism. Rather, it is a way to prevent records and CDs from going home with the myriad of individuals who will be rummaging through the collection, even mixing it with their own while they are on air. It makes sense. And it makes them less stealable. You can let the record stores in your area know not to buy material with your call letters on them and to contact you if any shows up.

TIP

When WXOX gets donated collections, we write the owner's initials on the front of the album. This helps us remember their generosity and keeps their musical taste in rotation.

FIGURE 7-4: A record from the WHRC collection.

© Jennifer Waits, Co-Founder Radio Survivor.

Developing a Digital Library

Some LPFM radio stations are nothing more than a digital music collection on auto-play. In my opinion, this approach to broadcasting runs contrary to the entire concept behind the Low Power movement and takes the "Community" out of the *Local Community Radio Act.* When there are no local voices being broadcast and no local issues being discussed, these stations become just another algorithm.

I sincerely hope that if you take the time to read this book it is not simply so that you can put your station on shuffle and then walk away from the controls. But it is possible, so bear my following opinion in mind: Automation sounds like the death of humanity and it is. When we let computers select the music we listen to, we lose the connections, the meanings, the emotions, and the threads of life that weave our collective experience together.

Still, even the stations that are most dedicated to having live programming will find themselves choosing between dead air and computer automation at some point. Before the IT revolution, we had to shut down the transmitter and go off air when no one was in studio. Today, we can set our computer to play music while no one is there. But this is dangerous. Software automation systems can develop playlists based on particular genres, announce the song titles, tell the time, predict the weather, and make all the necessary station IDs (see Chapter 10). With this kind of technology, who needs DJs at all? It is far too easy to push shuffle and walk away. Much easier than dealing with the infinite problems that arise when working with real live humans — but not nearly as much fun!

TECHNICAL STUFF

It is common for community stations to refer to their automation computer as DJ H.A.L. after the computer that guides the spacecraft *Discover One* deep into outer space in Stanley Kubrick's movie *2001: A Space Odyssey*. Over the years the DJs of WXOX have witnessed DJ H.A.L. take on a life of his own. He makes pretty good selections these days and even gets his own fan mail. We know he is our enemy, but he has tricked us into liking him. His secret? A great digital music collection.

Seeking digital selections

As an FM radio station, you will be able to get digital copies of just about any music you want, and often you will be able to get it for free. It takes a little work reaching out to musicians, labels, and promoters, but if someone at your station is willing to put the time into it, this can really pay off by exposing DJs to all kinds of new music.

Sometimes labels will want to follow up with you about how the music is doing in your rotation, which is fine, but don't let them pressure you into playing certain music to stay on their mailing list. You are in control of your airwaves and they should let you have the promotional music, no strings attached.

These days many performers are managing their own digital sales through sites such as Bandcamp (`bandcamp.com`) and Community Music (`communitymusic.net`). These avenues provide direct contact between musicians and station managers. Send them a direct message through these platforms or social media. Let them know about your station and there is a good chance they will let you have all the digital files you want to play.

Managing digital files

After you get digital music files, what do you do with them?

You will certainly want to keep these digital music files in your radio station computers. This sounds simple, but when you start thinking about it, it gets tricky. You have to make some decisions regarding the station computers — how many will there be, what will you put on them, and who will have access? Although there may be some computers set up for the DJs to access music, search online, and archive their shows, keep access to the more critical station computers tightly controlled — such as those performing automation, bookkeeping, and membership management.

For us at WXOX it seems too dangerous to allow our 100+ volunteers into the brains of DJ H.A.L., our main automation computer (see earlier in this chapter for an introduction to DJ H.A.L.). As a result, we have consciously chosen to keep this computer locked down. The DJs may use it to look at bookmarked pages but otherwise they are not allowed to touch it.

Because H.A.L. is connected to our soundboard, we keep the music inside his library strictly instrumental and foreign language. We need to be able to trust that this computer will never cuss or act outside of FCC rules (see Chapter 10 for more about FCC regulations). For us, it is important that H.A.L.'s playlist also provides a comforting listening experience and that the songs transition peacefully into one another. Therefore, harsh and jarring songs are excluded from the playlist. There is a fine art to developing an automated playlist, so make sure that you have the right person curating.

In terms of building a larger digital music collection that is accessible to your staff, I recommend using another computer altogether. This machine can include all kinds of songs, local music, and promotional releases. This secondary computer should be accessible to your DJs and may or may not be connected to your soundboard. If it is not, it can be housed in the listening room, acting as an offline library where your staff can preview the music and download content for on-air use.

I recommend that DJs create a *clean music library* on their personal computers that contains music that is 100 percent free of explicative and parental warnings. I also recommend that stations do this as well. You can simply throw all the music you know is 100 percent clean into this list and start to build your on-air library. If you are in a jam and need to throw something on to fill the silence, you have this list as a quick go-to. From there, playlists may include "In-Studio Performances" where you keep an archive of all your original recordings. You will also want a "Local Music" playlist. Others may be categorized by genre: "Jazz," "Classical," "Hip-Hop," and so on. Assign staff members as genre music directors (such as "Blues music director") to curate these specific categories and work on keeping the collection they're responsible for clean and current (find out more about station staff positions in Chapter 12).

TIP

Make sure you have a back-up (or three) of your digital music collection on an external hard drive. When so many people have access to your computers, you never know what may happen!

Some stations have found ways to share their digital music library online so that DJs can access remotely and preview at home. A program called Subsonic (`subsonic.org`) allows listeners to browse, play, and download the songs within your station computer and through any device that is connected to the web. Disseminating downloadable audio files can also be done through online sharing systems such as Google Drive, i-Cloud, or Dropbox. Be prepared because any sizeable library will require a financial investment in online storage space.

WARNING

To avoid copyright issues, stations that share audio files with their DJs should have a clear policy that any copyrighted music will not be further shared or played beyond what is necessary for the production of their radio program.

Using software to keep the music going

You will need a program on your computer to host your music library. Most computers come with software that can create playlists, shuffle music, and so on. but you will also need the ability to schedule official announcements. At the very least, this software will need to play your legal ID (see Chapter 10) at the top of the hour.

A number of automation software choices are suitable for station use. At WXOX, we have been happy with MegaSeg Pro. It is reliable, affordable, and user friendly but is only available for Macs. Some other automation systems include Station Playlist, RadioDJ, Rivendell, LibreTime, and Sam Broadcaster Pro. Table 7-1 shows a comparison chart of automation software commonly used by LPFM stations (thanks to David Klann of Broadcast Tool & Die for compiling this data); I include a portion of it here to help you make your music software decision. An extended and up-to-date list can be found online at `https://tinyurl.com/BTD-automation-spreadsheet`.

TABLE 7-1: Music Automation System Comparison Chart Complied by David Klann of Broadcast Tool & Die

	LibreTime	Megaseg Pro	RadioDJ	Rivendell	Station Playlist
Author	`libretime.org`	Megaseg	RadioDJ	Paravell Systems	Station Playlist
Runs on (operating system)	Linux, Cloud	MacOS	Windows	Linux	Windows, Cloud, Linux (with WINE)
Initial cost (U.S. dollars)	$0.00	$199.00	$0.00	$0.00	$286.00
Recurring licensing costs	$0.00	$0.00	$0.00	$0.00	$59.00
Program simultaneous "streams"?	Yes (with customization)	No	No	Unlimited	Up to 5
Technical support options	Email, forums	Phone, email	Forums	Phone, email, remote access, 3rd party support contracts	Email
Import logs/ playlists from external scheduler app?	No	No	Yes, M3U file	CounterPoint Traffic, Music 1, NaturalLog, PowerGold Music Scheduling, `Radio Traffic.com`, Rivendell Standard Import, Visual Traffic, WideOrbit Traffic, Custom	Natural Log, AdMaster, The Traffic Light, Radio Traffic, Radio Works, Music 1, Power Gold, RCS Selector, Natural Music
Sound Exchange reporting?	Yes	Yes	No	Yes	Yes
Scripting or macro support?	Limited	No	Limited	Yes (macro language)	No

	LibreTime	Megaseg Pro	RadioDJ	Rivendell	Station Playlist
Web integration?	Good	No	Limited	Somewhat, and improves with each release	Good
Common praise	Web-based, ease of use	Ease of use, well integrated with Mac	Ease of use, easy to get started	High-end at a low entry cost, can do just about anything	Ease of use, easy to set up and get going
Common criticisms	Not really a full-featured radio automation system	Apple only	Lacks more advanced automation features	Lack of good, complete documentation & complex to set up	Reoccurring fees

Archiving Original In-House Productions

As a community radio station, you will have so much talent coming through your doors, and you must be prepared to keep up! Maintain an archive of the interviews and live performances that occur on your FM broadcast. With the proper settings on your recording devices, you can get a higher-quality recording in-house than you can by recording via the Internet.

Provided that you have all the proper performer releases (see Chapter 11), you will be able to play and redistribute the original productions created on your broadcast. This will be great material to draw on in a jam and it provides unique listening experiences that sets your station apart.

TIP

Radio stations often produce albums of their live in-studio performances and these can be excellent fundraisers. For example, for 25 years KFJC Los Altos has been producing *Live from the Devil's Triangle* (an annual compilation of in-house performances), and WFHB Bloomington produces an album from their annual Lotus Blossom Festival.

Creating an audio library of original material is a fun project for any radio station. It can include anything from sound drops to sound effects. The voice recorder app that comes on most smartphones is a remarkably good way to capture sounds on the go. It is easy to use and a great way to record Grandpa George's limericks at the family reunion or the call of the whippoorwill deep in the Georgia woods.

For those seeking higher audio quality, there are a number of affordable portable recording options made by Tascam, Sony, and Roland. Our station keeps two Zoom hand-held recorders available for our DJs to check out.

TIP

Audio professionals know tricks that can help produce a cleaner audio recording. Perhaps there is someone in your community that can lead a field recording workshop for your staff? Remember to always wear your headphones when recording so that you can listen for any interfering sounds. When you get the basics down, the world of sound will open up and you will discover that there is never a silent moment!

IN THIS CHAPTER

» Financing your operations

» Cultivating support and funding

» Building incentive programs and sales

» Developing fundraising events

Chapter 8

Going for the Gold: Developing Station Income

Truth be told, I am a believer in the Field of Dreams. If you build it, they will come. Of course, building the dream includes developing ways to keep it financially afloat and discovering ways to fund your dream-building project!

As noncommercial stations, LPFMs are restricted from advertising (thankfully) but many other ways exist to generate income for your community radio operation. This chapter discusses some unique strategies for fundraising that are well within Federal Communications Commission (FCC) guidelines (find out more about the FCC in Chapter 10). Sometimes you have to get creative, so this chapter is here to help!

Making Friends and Creating Opportunities

As soon as you have made the decision to build a radio station, the first thing you need to do is get your boots on the ground. Start talking to folks, finding out who in your community may be interested in working on this station, helping it become a reality. Radio is exciting and folks may get enthusiastic very quickly! In addition to talent and technologists, you will need to find some investors.

TIP

Think about like-minded businesses and organizations that you can connect with in your community. As an arts-focused radio station, ART FM started by contacting leaders in local creative circles right away. We reached out to administrators at local nonprofits by finding their emails on their websites. We wrote asking if they could meet for coffee to talk about our new radio idea. Almost everyone was enthusiastic. Many wanted to participate. Many were able to offer direct support. This included equipment, studio space, community partnerships, and programming ideas.

REMEMBER

Keep in mind when fundraising that you are not asking folks to give, you are offering them an opportunity to invest in their community — emergency response services, technical training courses, increased communication among citizens, and more! Showing potential investors that building a radio station is a lot more than playing records is such an important thing to do.

After you begin getting donations, you will need to keep track of them carefully. The online resources of this book provide both a donor spreadsheet for tracking donations and a budget workbook that helps you to keep up with all of your non-profit's financial matters. I discuss the importance of keeping everything transparent and on the level in Chapter 2.

Organizing a Capital Campaign

A *capital campaign* is an intense push by an organization to raise a certain amount of money within a specific time frame. Starting a radio station is an ideal time for a capital campaign; after you have your LPFM license (see Chapter 3) you have a limited amount of time to purchase your transmission equipment and get up and running.

REMEMBER

To get started with your capital campaign, set a monetary goal and a time frame in which you want to raise it. Then you need a platform for raising the money. Of course, donors can always give direct cash donations or send checks in the mail but some of the most effective fundraising is done online. As soon as your website goes live, set up a "Donate" button on your home page. If you use PayPal (`paypal.com`) or Venmo (`venmo.com`) you will be able to get a button code to embed on your site through their admin tools.

Some radio stations have extended fundraising campaigns that involve weeks of special broadcasts. At ART FM we use our community-wide day of giving to get it all done in a 24-hour time period. Although we encourage donations on this one specific day, we spend many weeks building publicity for the event.

TIP

Crowdfunding sites are very popular ways to raise money. If you are a Federal 501(c)3 tax-exempt organization that wants to go down this route, be sure to register with a site that allows your donors to make tax deductions, such as Indiegogo (`indiegogo.com`), GoFundMe (`gofundme.com`), FundRazr (`fundrazr.com`), or Mightycause (`mightycause.com`).

WARNING

Take some time reviewing crowdfunding sites to see which one has the features you desire. First and foremost, find out if there is a fee to use the platform and if there are transaction fees involved. Watch out for sites that automatically add donations for themselves. GoFundMe, for example, charges a 3 percent transaction fee. This platform contribution is technically optional but if your donors are not careful, it automatically adds a self-serving tip. This can be a real turnoff to your patrons when they accidentally give more than expected.

Some crowdfunding sites offer the opportunity for fundraisers to set up individual giving pages or peer-to-peer fundraising pages. These pages can be customized by station volunteers for personal asks, yet all of the donations go directly to the nonprofit organization. I am a big fan of these peer-to-peer pages as they add a personal touch to the fundraising. It gives the DJs a higher stake in the game and it makes their acquaintances more likely to donate. Awarding internal prizes for the fundraiser who brings in the most donors or to the most creative page helps motivate your staff to get involved with "The Ask." Promise a big party when you make your goal or have one along the way to help you get there!

TIP

As the money comes rolling in, find a fun way to let your supporters know how the campaign is going. Some radio stations get very creative with this. For example, instead of using the typical fundraising thermometer, broadcasters tend to go with a *Money-O-Meter*. This is a graphic based on an audio meter that goes higher and higher with each new donation. These can be made digitally or handcrafted. As a newly licensed LPFM, Radio Tacoma WOWD ran a brilliant fundraising campaign by creating cardboard models of its dream equipment and inviting the public to invest in the reality. In Figure 8-1, you can see a photograph of one of their cardboard computers that was used to solicit donations.

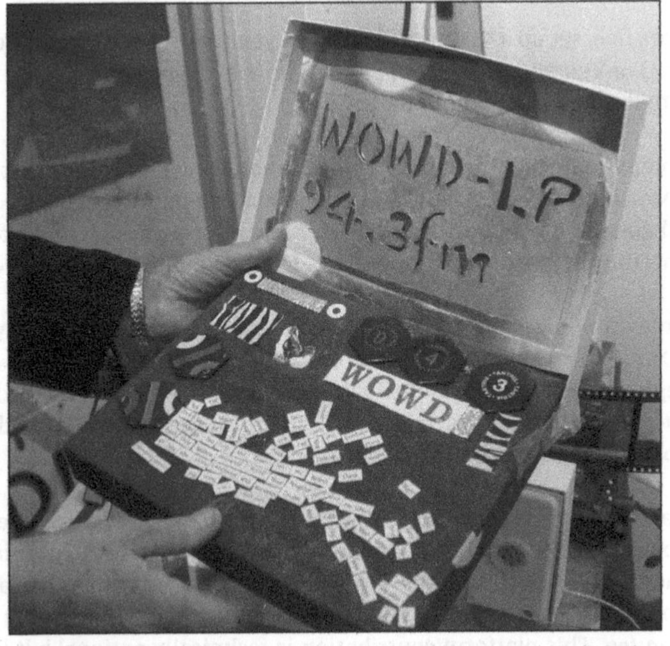

FIGURE 8-1: Prior to going on air, Radio Tacoma WOWD created their entire studio out of cardboard and invited potential donors into the station to imagine how it might look with some community investment.

© *Jennifer Waits, Co-Founder Radio Survivor.*

COMMUNITY GIVING DAYS

Here in Derby City, we like gambling a lot more than we like asking for money. Therefore, *community giving days* are right up our alley! During these 24 hours of giving, area nonprofits compete for cash prizes and matching grants provided by charitable community foundations that encourage local citizens to give to their favorite nonprofits, knowing that on this day, each donation will go a little further. These events come under different names but they all seem very effective.

For example, thousands of dollars in prizes are offered during the *Give for Good Louisville* community giving day, such as the *Up All Night Prize* for the most unique donors between 2 a.m. and 4 a.m. and the *Far and Wide Prize* for donations from all 50 states. We like to win them all!

Having an FM frequency to amplify the call really helps — we use it to stay on air with live DJs for the complete 24 hours, asking for donations after every song or two. This, accompanied with a serious online social media campaign, physical mailings, and in-person party helps us reel in a number of bonus prizes. By the close of the contest, we have donations from around the country and raise the majority of our entire operating budget in a single day!

A great thing about this program being run through the Community Foundation of Louisville is that in addition to generating a lot of publicity for local nonprofits, they handle many of the business aspects of fundraising including generating donor spreadsheets and tax letters. This leaves you to simply send a sticker and a cute thank-you note in the mail!

Building Station Membership

Noncommercial radio stations often run an annual or semiannual membership drive. A *membership drive* involves using the airwaves to ask listeners for their financial support.

On-air membership drives can be very effective and there are many stations that depend on these annual fundraisers alone. The key is to make them fun and fairly quick interruptions to regular programming. Don't let them drag on forever or listeners will just change the channel.

In addition to soliciting station donations on air, membership drives can be promoted online and through direct mail campaigns. Stations often offer special programming at this time that encourages folks to tune in — perhaps an alumni show, live performance, or a trivia program with local celebrities involved. WFMU General Manager Ken Freeman is the king of the pledge drive stunts. He will do almost anything to help his station meet its goal including tying helium balloons to his lawn chair and floating high above their studios in Hoboken (do not try this at home!).

Fun Fact: The membership drive, now so perfected by NPR, actually originated in community radio with Lewis "Lew" Hill, founder of the Pacifica Network. Find out more about Hill and his ideas on station development in Chapter 17.

Station membership can mean something different at each station. You can do whatever you want to do in terms of developing membership levels, incentives, and so on. And you can change strategy whenever you feel necessary.

For example, ART FM has a few membership levels. Basic membership costs $97.10. We couldn't decide on an introductory price so we just went with the number of our frequency. We also began offering a half-price artist discount to anyone who created a 2-D piece of art and sent it in our direction. Photograph, painting, drawing on a napkin, whatever! Everyone is an artist and everything is art! We keep some of the artwork for our station collection but we don't have room for it all, so we put some pieces in our annual silent auction as a means for raising additional funds for the station. You can read more about hosting silent auctions later in this chapter.

In addition to basic membership, it can be nice to offer advanced levels that recognize those who really do a lot for the station. Sometimes people give in ways other than money and it's nice to recognize them here as well. Your landlord, for example, may donate space; your web designer may donate hours of work. These advanced-level members may enjoy special events and backstage tours. Try to involve them as much as you possibly can! After they have given, send each donor a thank-you card personalized by one of your DJs. In addition, why not send out an e-newsletter, invites to members-only events, discounts on tickets, offers to read their names out on air, or name them on your website as a station donor.

TIP

At ARTX FM, we've been told that three membership levels is an effective number, but we have six. You can see these on our website: `artxfm.com/donate`.

REMEMBER

After receiving a donation, a station should thank their donors immediately with an email and/or a handwritten note. 501(c)3 organizations should be sure to send a letter from the station that includes the amount of the donation and your organization's EIN number so donors can document the deduction. Some nonprofits send these letters right away, others at the end of the year, and still others just ahead of the tax-filing deadline. It is up to you but staying on top of this will be greatly appreciated by your donors.

If you offer any physical items with membership, the value of that item will need to be deducted from the value of the donation. Say, for example, if you offer T-shirts valued at $20 as an incentive for donors who give $100 or more, they will only be able to deduct $80 from their taxes. For this reason, at ART FM we do not offer physical incentives to our donors but instead the reassurance that every dollar they give goes directly into our noncommercial community radio programming.

WXOX has offered "2 for 1" sweetheart memberships in the past. As you can see in Figure 8-2, we made Valentines that folks could send to their loved ones, letting them know they had received a gift membership. Many of those gifted memberships continued to support the station after their first complimentary year ran out.

At WXOX we ask that all of our program hosts become members of the station and it really helps us stay afloat independently. We let them know that their donations go directly into covering the essentials such as utilities, insurance, music licensing, and equipment updates.

Station memberships can be paid all at once or monthly. This is very easy to set-up through PayPal or online banking. Many stations are moving to monthly donor payments, which is great because it spreads donations out across the year and renews automatically. In other words, you only have to ask once!

FIGURE 8-2: Valentine's Day "2 for 1" membership used at WXOX.

© Sharon M. Scott (Book Author).

Developing Partnerships and Sponsorships

Partnerships are like memberships (see the previous section for more on membership) but for businesses. At ART FM we reserve partnerships for businesses that have helped us in a big way. Seiller Waterman (`derbycitylaw.com`; the attorneys who helped us get established) and Occam Design (`occamdesign.com`; the company that provides our tower site), for instance, are station partners. We recognize our partners a few times a week on air. Their logo goes on our website, studio wall, and any printed publications. They get surprise swag throughout the year and VIP tickets to any station event.

Sponsors are the businesses that provide food, beverages, supplies, and other items or services to station events. At ART FM, we thank our sponsors on air in exchange for their donation. They also get their logo posted at the events they sponsor and in any related publications. Their name is mentioned on air and on stage. Like the station partners, they get VIP tickets and signage opportunities at the sponsored event.

As with memberships, radio stations can feel free to develop their own partnership and sponsorship programs that work best for each organization. It is okay to adjust these programs as you move along, so long as you uphold your legal obligations (see Chapter 10).

Working with Underwriters

Underwriting offers a unique opportunity for noncommercial radio stations to accept money from business or individuals in exchange for on-air recognition. You will often hear noncommercial stations saying "This program is brought to you by . . . ." This statement will be followed by the name of a business and its contact information. It may include facts about the business but qualitative or persuasive language cannot be used.

Although FCC law does not allow LPFM stations to advertise for-profit businesses, it does allow them to thank the businesses that support them on air. In fact, it requires that they do so (for more on FCC laws, see Chapter 10). Factual information about the sponsor is allowed so long as it does not endorse, incite, or encourage anyone to buy. Therefore, never mention prices on air, and never issue a call to action on air.

Quantitative words are allowed in underwriting announcements but qualitative words are not. Do you remember those tests in grade school where you had to eliminate all the adjectives? Well, you are playing that same game here — so avoid language such as "good," "amazing," and "the best." Underwriting announcements must stick to the facts.

After you accept money from an underwriter, you can be in very serious trouble with the FCC if you in any way endorse their for-profit business, events, or products. I have seen a number of stations get busted for this in recent years. Fines are in the neighborhood of $10,000 to $30,000. Review FCC law by heading over to Chapter 10 for more information about what you can and can't say on air. Underwriting announcements must adhere to all the rules.

Underwriting has many advantages over advertising. First and foremost, it builds a common bond between listener and business and shows them that they are both supporting the same community cause. Rather than being forced to hear advertisements interrupting the music, the listener grows an affinity for the business that is actually making the programming happen. Underwriting is less intrusive on the programming than advertising and will not irritate the listener or interrupt the mood in the same jostling way.

Be careful about what businesses you have underwriting your shows. Their ethics will become a reflection of yours. The FCC has ruled that broadcasters are not "common carriers" and therefore not obligated to afford universal access. The broadcaster has the liberty to determine which underwriting and which programming they accept or decline.

WARNING

There are, however, a few caveats. Noncommercial stations may not air underwriting announcements from political parties or tobacco companies. And until federally legalized, LPFM stations with federal licenses, cannot accept underwriting support from the cannabis industry. Remember the following advice:

» Underwriting announcements may include
  - Location and contact information of the underwriter
  - Slogans, brands, and trade names that identify and do not promote
  - Value-neutral descriptions of products and services

» Underwriting announcements may not
  - Promote any for-profit service, facility, or product by way of endorsement, calls to action, or special inducements to buy
  - Advocate the views of a given person on a matter of public interest
  - Endorse a specific candidate for office

UNDERWRITING EXAMPLES

Writing good underwriting copy can be tricky. Here are a few examples of underwriting announcements that stick to the facts and do not violate any FCC rules.

Example 1: "This portion of programming is brought to you by Nachbar located at 969 Charles Street presenting American noise rock duo Lightning Bolt on Monday, March 26th at 8 p.m. This is an 18 and over show."

Example 2: "Support for WXOX comes from Louisville Community Acupuncture, offering acupuncture in a community setting and in private rooms for the past 8 years. More information available at Louisville Community Acupuncture dot com."

Example 3: "This program is proudly sponsored by Rainbow Blossom Natural Food Markets. Family owned and operated since 1977, Rainbow Blossom offers 100s of locally sourced products including organic fruits and vegetables. Details at Rainbow Blossom dot com."

Finding Government Money

Local, state, and national government agencies all offer money for nonprofit and community development projects. It will take some reaching out to find it. Connect with your local department of neighborhoods and mayor's office. Make sure you are on all the right email lists. Talk to your city councilperson. They often have some discretionary money for community building that would be perfect for developing a neighborhood radio station! Likewise, there are public safety grants that can finance items like your Emergency Alert System (EAS). Use the Internet to search for your state emergency management agency. Their website will have a list of county contacts you can reach out to for assistance.

A number of state and federal grants exist for start-ups run by women and minority groups. This is all the more reason to embrace diversity on your board! There are also economic development grants to jump-start businesses in certain geographic areas. Community health incentives may give you money to read important public service announcements. Some LPFMs, for example, got funding from hospitals by sharing information updates during the COVID-19 pandemic. The money is out there, but you'll have to go find it!

TIP

Most cities have community foundations and nonprofit centers that connect donors with local charities. They may also provide seminars and training on nonprofit development. Introduce yourself to these agencies right away; see if they have any workshops or if they can recommend avenues for financing your new community station. Be sure to remind them what an asset your station can be to all local nonprofits, and they may be able to make some valuable introductions.

Grants that are aimed at specific programs seem to get more traction than general asks. For example, requests seeking support for a teen radio program may have more impact than asking for help with general operating expenses. Be aware that some grants will ask what percentage of the board donates to the organization with the expectation that all of them will do so. It makes sense that a granting agency may not want to invest in a board that does not invest in itself.

For more guidance on grant writing, I encourage you to look at *Grant Writing For Dummies*, 7th Edition, by Beverly A. Browning (John Wiley & Sons, Inc., 2022).

Considering Incentive Programs

One easy way to bring in extra dollars for your tax-exempt organization is to sign up for incentive programs from businesses that give back to your charity whenever you send them customers. Although these businesses give your organization percentages of each purchase, there's a double-edged sword when these incentives are offered by larger corporations such as Kroger and Krispy Kreme, as your community station really should be a friend of local businesses.

Try to seek out local restaurants that offer community kickback nights in which they provide a cut of their sales for your nonprofit organization. These events can be fun ways to get your staff and listeners together.

In both these instances, you will be able to promote however you want in print and online, but be careful about how you get the word out on air. Since your organization will be splitting the sales with a for-profit venture, you will not be able to promote sales or encourage people to show up. Do not mention prices or the percentage kickback.

Driving Vehicle Donations

Fire up your engines and get moving on this, because vehicle donations are a fantastic way to raise money for your organization! A *vehicle donation* program allows listeners to donate their car, boat, or motorcycle to your organization through a third-party service that handles all of the logistics, including the pickup and the sale of the vehicle. After deducting their expenses and a small service fee, the company donates the remainder of the income to the radio station.

This is one of my favorite fundraisers because after registering your nonprofit organization with a vehicle donation company, you basically don't have to do anything at all — besides getting the word out, you just watch the money roll in!

At WXOX we've had good luck with CARS (careasy.org) but there are several companies that do this work on behalf of nonprofits such as Cars2Charities (cars2charities.org) and Charity Cars (800charitycars.org). With CARS, we were given a landing page and a unique phone number.

Everyone I know who has donated a vehicle to our station has felt like they got a good price for their car and were pleased to make a contribution to the radio station. The service provides them with letter documenting their tax-deductible gift, and the radio station celebrates them with a thank-you card and full membership privileges.

Maximizing Merch Sales

Work that logo! Print it on T-shirts, hoodies, baseball caps, record bags, and more! It all starts with great art work and a brand people want to represent — then you can sell anything with the logo on it!

WXOX has done very well with merch sales but we know that it could even get better if we had someone willing to get it out there and push it. There is a good headroom for marking up T-shirts, hats, and hoodies. I recommend starting there. You will have a built-in audience of DJs and family members. These items also sell very well at concerts, open houses, and community events. They also make nice giveaways. And there's no better advertising than someone walking around in your station's T-shirt! (Except maybe driving around with a bumper sticker, which should always be given away for free!)

Perhaps you can get a local store to carry your merch and provide some airtime in exchange for the floorspace. For example, "WXOX merch is now available at Sugar Town Vintage located at 205 S. Shelby Street in downtown Louisville." This creates a win-win situation for everyone involved and your merch is getting out there onto the streets!

Hosting Fundraising Events

Events are great ways to raise money and energy for your station. It's always so much fun to get our DJs and listeners together in person. There are so many great event ideas for nonprofits — the sky is truly the limit!

One bonus about working in radio is that you have your own built-in talent! It shouldn't be hard to find staff members who are willing to spin their records live. As a radio station, it is also easy to book the best bands in town, so musical events make a lot of sense. Many bands are willing to share their talent for the community radio cause. In exchange, they get promotions on air, online, and in print materials. Place their music in studio for the DJs to play during the weeks ahead of the benefit. When possible, live broadcast from the show so listeners worldwide can join in the fun!

I recommend selecting a visual artist to create a collectable poster for the event (check out Figure 8-3 for an example from Dave Huckins of Yondernod). Make a postcard version that you can send to supporters in the mail, hand out to people on the street, and so on. For not too much of an expense you can put the image on the front of the event tickets through a printing service such as My Zone (`myzone.com`) or Avery (`avery.com`).

FIGURE 8-3: Collectable poster and tickets to the WXOX Voix de Ville. Poster by Dave Huckins.

© Maxwell Cash Selby.

REMEMBER

Events and fundraisers can be a lot of work but they are great team builders and revenue generators for your station. DJs seem to enjoy collaborating and getting to know one another because it is usually a brief passing when they cross each other at the station. Plus, it is great for them to connect with the community and hear directly from the listeners. In the planning and organization of station benefits, you will find out who is dedicated to building your organization and who you can rely on in a jam. Friendships will be made while problems are solved — and you will emerge a better, stronger station!

Events may take place at your studio or in another location. Often the person or organization providing the space will want to be added to your event insurance policy as an additional insured party. ART FM has found that our professional liability covers events in other places and that the additional insured party is added at no fee but you should check the specifics with your provider.

To put on a good show, you need good sound equipment. If you don't have your own PA system yet, rent one for the night from a local music shop. These rental companies often employ sound technicians and may insist that you hire one of them to manage the equipment. If you don't have experience or a sound person that you trust, this can be very helpful. If things are getting expensive, ask if the rental company is willing to make a trade for underwriting or become an event sponsor. There is a good chance that you can at least negotiate a discount!

Nonprofits can really bring in quite a bit of money at an event with a silent auction. A *silent auction* is a fundraising event in which participants try to outbid one another for prizes before the event officially closes. Nonprofits often ask local business to donate items and gift certificates to the auction. When the event is over the supporter goes home with a great prize, the business gets some excellent publicity, and the nonprofit organization has raised some amazing funds. In most cases, everyone has a great time doing it!

One year, WXOX had a surplus of misprinted record bags so we gave them to our radio hosts and asked them to fill each bag with at least three things that represented their show. These became auction items on which the patrons would bid. The creativity with which the DJs responded to this project was astounding! They showed up with music, art, and collectables. They brought in gift certificates, musical instruments, and a host of libations. The listeners enjoyed bidding on the record bags and winning unique treasures from their favorite shows. These silent auction items earned us nearly $10,000 the very first time we tried it!

Many silent auctions are now held virtually through sites such as 32auctions (`32auctions.com`), OneCause (`onecause.com`), and GiveButter (`givebutter.com`). This can be a good way to let distant supporters participate in the auction, but remember that you have to give the site facilitator a cut of the revenue. Price it out and decide if this is worth it to your organization.

WARNING

A quick word about raffles: Check your state and municipal laws as raffles and lotteries are considered a form of gambling. You may need a special license. If this is an expense and bureaucratic hassle that you would prefer to avoid, why not place these enticing items in a silent auction instead?

IN THIS CHAPTER

» Paying for utilities, insurance, and inspections

» Saving up for new technology and music licensing

» Factoring in staff and development costs

» Budgeting for professional costs and communications

» Ensuring that your financial records are watertight

Chapter 9

Holding the Line: Managing Station Expenses

It takes money to maintain a radio station. That's no joke.

However, it can be done and it can be done on a shoestring. Many expenses will come your way, but do not be intimidated. And don't get yourself in debt. You can take it as it comes, work within your budget, and build your way up to the top. A community radio station is a rare and beautiful thing.

This chapter helps you get a grasp on the annual operating expenses involved with maintaining a LPFM radio station after you are up and running.

Maintaining Your Home-Base

Electricity, water, and gas bills depend on the size, location, and composition of your building. Be sure to consider all these variables when selecting your studio location.

Insurance is expensive; utilities are expensive; and rent always seems too high! A good bit of your budget will go to these items and there are not a lot of ways around that. For a few insurance tips, see Chapter 2 where I discuss the essentials for nonprofits. At a minimum, you will want business liability and equipment coverage, but you may also want broadcaster's insurance to protect the station from lawsuits when DJs say stupid things on air, as well as a policy that personally protects your board of directors.

It is possible to find donated space for your studio; think about businesses that may have extra room available and could benefit from sponsorship announcements on your broadcast. There's a long history of hotels hosting radio stations. KPSQ (kpsq.org), for example, is located in Room 310 of the Graduate Fayetteville Hotel near the University of Arkansas. The hotel has donated the studio and antenna tower space in exchange for on-air shout-outs: "This is KPSQ, broadcasting live from the Graduate Hotel in downtown Fayetteville." What a perfect arrangement! In addition to hotels, studio warehouses, museums, and libraries can all benefit from housing community radio stations — and just imagine the collaborative programming! Find more on this in Chapter 8.

You may want a P.O. Box for your organization and this currently runs between $100 and $150 annually. A small box will usually work fine because occasional larger packages are often placed in another lockbox for pick-up.

If your station is dedicated to providing emergency services, you want a landline telephone to stay connected with local citizens and first responders. Landlines cost around $100 a month, so it is no minor expense. It is possible to get a free Google Voice (voice.google.com) phone number for your *request line* (a public number where listeners ask the DJ to play specific songs), but if the Internet goes down so will your access to any Internet-based line and the critical ability to take calls during an emergency.

Check with your city and utility company to see which site inspections may be required in your area. In Louisville, businesses — even nonprofits — have to get their fire extinguishers and backflow prevention devices inspected annually. These are minor expenses — so long as you pass all the tests!

Factoring in The Price of Technology

This is the modern world — so you will need an IT budget. The following sections discuss the main IT expenses you need to consider.

Internet service

Having solid IT networks is very important to your operation. A stable webcast depends on a consistent upload speed. However, audio streaming does not take that much bandwidth, so you should only need the basic service (budget for around $100 per month).

TIP

After plenty of headaches from service outages at the most inopportune times, ART FM ultimately decided to go with two separate Internet providers. One is dedicated to getting our studio audio to the FM transmitter site by way of an audio codec (see Chapter 4); the other is used to generate the web stream (see Chapter 5) and the station's WiFi. The use of two Internet providers allows us to be on air — at least in some capacity — in the event that one of them goes down. We prioritize the FM signal but can always flip-flop the connections if necessary.

Webhosting

For your website, expect to pay hosting fees of around $200 per year. You will need to maintain your domain name registration for $10 to $20 per year. If you web stream, you need to pay this host service as well, and this will cost another $150 to $200 per year.

TIP

All in all, budget somewhere between $350 and $500 per year on website-hosting costs.

Remote access

You need a way to log in and control your computers remotely. This convenience can range in price from the free-to-use Chrome Remote Desktop (`remotedesktop.google.com`) to upward of $800 annually with programs such as LogMeIn (`logmein.com`). Having the ability to access your computers remotely is super important, to prevent you from driving to the station at inconvenient hours to handle broadcast problems. It also allows you to go on vacation and get work done on your station computer while you relax. To me, this convenience is worth a premium price!

Video conferencing

During the COVID-19 pandemic, most stations found the need to get some kind of online conferencing platform up and running in order to hold meetings, run training, and create social events. This convenience of meeting remotely seems to be one thing that will stick around even in healthier times. Zoom costs around $200 per year, but Discord (`discord.com`) and Google Meet (`meet.google.com`) offer free-to-use alternatives.

Licensing Music

If you want to play music on your LPFM station, you need to get permission from the performers and songwriters to do so. Because it would be impossible to contact all the individuals necessary for this sort of permission when broadcasting 24/7, radio stations instead pay annual fees for blanket licenses from Performing Rights Organizations (PROs) to play the music in their catalog.

Music licensing seems intimidating to some but it is actually quite simple. You register, you pay, you play. And then you get a cool sticker to put on your window!

REMEMBER

Two types of licenses for radio stations are available: one for FM broadcasting and another for streaming online. If your station does both, you will need licenses for each of them. You can find out about the rules for implementing these licenses in Chapter 10.

FM transmission

The easiest, most affordable way to get permission to broadcast copyrighted music on your station is to apply for a *blanket license* (or *statutory license*) from the Performing Rights Organizations (PROs). The Big Three are American Society of Composers, Authors, and Publishers or ASCAP (`ascap.com`), Broadcast Music, Inc. or BMI (`bmi.com`), and Society for European Stage Authors and Composers or SESAC (`sesac.com`). Each of these agencies license the music belonging to different copyright holders — the songwriters, composers, and publishers. Most of the musicians who have their music copyrighted in the United States will be covered by these three organizations.

For online broadcasters, the *Digital Millennium Copyright Act* (DMCA) places limitations on how often you can play certain music. See Chapter 10 for more specific information.

To get a blanket license for your station, contact the PROs directly and let them know when you wish to begin playing their music. Let them know that you are a nonprofit and that you are an LPFM. They may have special rates. Registering can be done in a few easy steps online or through a quick phone call or online application. No problem! If you webcast, be sure to let the PROs know as well; BMI and SESAC will have you set up a separate account for the streaming fees, while ASCAP will combine it with your FM license.

TIP

Use the following contact information to register for your blanket FM licenses:

- **ASCAP:** 1-800-99-ASCAP or licensing@ascap.com
- **BMI:** 1-800-925-8451 or bmi.com/licensing/contact
- **SESAC:** getalicense.sesac.com

Presently, all things combined, LPFM licensing fees cost approximately $1,000 a year for FM broadcasting and $1,500 for webcasting. They creep up a little higher after each year of service. It is a serious expense but considering that you may play music 24/7/365, it actually seems like a decent deal (and hopefully it is for the songwriters and performers as well!).

TECHNICAL STUFF

On their website, ASCAP poses the question "Where does the money go?" To which they answer, "Quite simply, to our members. All the fees we collect are distributed as royalties, after deducting operating expenses (currently 11.3%)."

REMEMBER

Brokers and third-party services will handle music licensing for you. Live 365 (live365.com), accuradio (accuradio.com), and last.fm (last.fm) can handle the streaming, reporting, and licensing all rolled into one package but do your research — after you start to accrue online listeners, their fees can become more expensive than simply paying the PROs directly.

Digital transmission

Stations that wish to play copyrighted music online must first file a Notice of Use of Sound Recording with the U.S. Copyright Office. This is a one-time filing with a fee of $40 that is sent to the Library of Congress Copyright Office where it will become public record (obtain your Notice of Use form here: copyright.gov/forms/m-269.pdf).

In addition to paying ASCAP, BMI, and SESAC (see the previous section) for both FM and online transmission, stations that stream online also need to pay annual licensing fees to SoundExchange (soundexchange.com). Unlike the other PROs that represent the songwriters, SoundExchange represents musicians and performers. It collects and distributes royalties on music that is broadcast digitally.

Noncommercial webcasters pay $1,000 annually to SoundExchange until they exceed 159,140 aggregate tuning hours (ATH) each month. *ATH* refers to the total number of cumulative hours listeners have spent listening to your station within a certain time frame. The number 159,140 roughly translates to 214 online listeners per hour, all day every day — which would be pretty extraordinary in the world of noncom broadcasting. To register your station with SoundExchange, visit their Licensee Direct portal: `licenseedirect.soundexchange.com`. Noncommercial stations report to SoundExchange quarterly with information about the music they have played on air. Stations that are licensed to accredited educational institutions can get a $100 reporting waiver as well as discounted membership fee.

There are a few options for submitting the Reports of Use to SoundExchange. You can upload the file to the SoundExchange Licensee Direct website (`licenseedirect.soundexchange.com`), but because it can be glitchy, you can also email your report as an attachment to `reports@soundexchange.com`, or you can send them a disc or USB drive by standard mail or special delivery. Hard copies of the reports are not accepted.

David Byrne and Martha Reeves sit on the SoundExchange board of directors, so hopefully they are working hard to make sure that this money actually gets to the musicians who deserve it!

TRACKING YOUR ON-AIR PLAYS

Spinitron (`spinitron.com`) or Pacifica's Confessor (`pacificanetwork.org/pacificainternetpackage`) are both excellent programs that will generate reports that you can submit directly to SoundExchange and anyone else that may request it.

There are fees for these services but as someone who once typed all of the playlists from the handwritten sheets DJs turned in each week, I can assure you that they are worth every penny! Both Spintron and Pacifica are sensitive to nonprofit budgets and the price will vary depending on how many add-on services you get.

Spinitron and Confessor both provide additional paid features such as program archiving and automatic song recognition. At WXOX we have loved the ability to embed the track lists on our website in real time so that people know what we are playing, when we are playing it. You can also connect these systems to your transmitter with a Radio Data System (RDS) that allows small bits of information such as song titles, artists names, and secret messages to be embedded within the FM transmission and appear on the digital display in automobiles. Pretty darn cool!

Hiring Personnel and Growing a Happy Team

Community radio is a labor of love. If you don't love it, don't do it. It will ruin your life. If you do love it, you will wake up every day ready for the adventure! It will be rewarding in so many ways. There are ways to make a living working in community radio, but if the money is your motivator, you missed the turn toward corporate radio a long way back. LPFM radio stations are nonprofit by nature and that's part of their beauty.

To avoid steep overheads, a community station depends on its volunteers. You will be surprised; after you get your station started, many people will want to help. Organizing volunteers takes a strong leader and someone who can motivate others and roll with the punches. Without the promise of a paycheck, be prepared for volunteers who disappear at the last second or fail to complete critical jobs.

REMEMBER

Every station needs someone it can depend on 24/7/365, and that person is the general manager. The general manager runs the day-to-day activities of the station and makes sure that it is acting in compliance with local, state, and federal law. They do what it takes to keep the organization moving forward. Likewise, the Federal Communications Commission (FCC) requires that every station have a chief operator who is responsible for making sure that the station is staying within the technical compliance of its license. The chief operator is also responsible for making sure the studio equipment is functioning properly, and is often the busiest person at the station.

The general manager and chief operator may or may not be the same person (they may even be you!). Either way, they need to get paid. I discuss these roles in more detail in Chapter 12. These jobs may start off as volunteer positions and then move into contractor positions before becoming actual employees, but it is important to find a paycheck for these essential individuals as soon as possible. The longevity of your station depends on being able to support key personnel.

Think about what other roles you need to fill. For example, if you need help with your website or graphic design, get that in your budget. The same goes for promoters and contractors. Often these individuals can be hired on a per-case basis rather than as full-time employees. Sometimes they will make an underwriting trade (see Chapters 8 and 10 for more on underwriting do's and don'ts). I recommend finding someone in your local community who is just getting started — see what kind of energy they can bring.

When it comes to website design, using template sites such as Wix (`wix.com`) and WordPress (`wordpress.com`) allows non-programmers to take control of their own websites as an alternative to developers who can charge high hourly rates just to make simple updates. These web-template sites offer ad-free plans that cost between $250 and $300 per year.

Finally, don't forget that someone will need to clean and maintain the studio space — inside and out — on a regular basis. You may need to hire a janitor, a painter, a gardener, and a snow shoveler — or you may be all of that rolled up into one! Radio is not always glamorous.

Most community radio stations operate with a volunteer on-air staff that keeps the broadcast raw, pure, and honest. As much as you appreciate your DJs, be wary about adding money to the situation. Not only would paying DJs to cover a 24-hour broadcast day break the bank of most young nonprofits, it would change the culture of your station. Instead, as your organization grows, consider adding paid positions for music directors, news directors, and training directors who are putting in hard work behind the scenes.

Showing staff and donor appreciation

You can never thank your volunteers enough. They are the heart and soul of your station, and they are the reason listeners are tuning in. The program hosts give so much time, energy, and creativity both on air and off that it so important to let them know how much you value everything they do. Taking time for this is not always easy with so many fires burning at all times, but it is always appreciated.

I will leave it to you to be creative here. Just be sure to work within your means. Keychains and patches are among the inexpensive items we have given as end of year gifts to the WXOX staff. More recently we were able to step it up to lunchboxes and ceramic mugs. We are presently working on a series of custom merit badges like those shown in Figure 9-1 that will honor WXOX volunteers for all their unique accomplishments.

Volunteer appreciation parties will really be helpful for station team-building. During the regular rotation, DJs can be passing ships as they move in and out of the control room. You will know them all and it is nice to give them a chance to get to know each other, too. A few boxes of pizza will go a long way on a tight budget.

The board may want to celebrate DJs with trophies, lapel pins, or certificates to recognize milestones such as program anniversaries or important volunteer work. We award DJs with "Golden Headphone" necklaces on their five-year anniversary — our WXOX celebs enjoy showing them off at important local events.

FIGURE 9-1: Custom WXOX patches created by DJ Shawn O'Donnell of *Power Trash.*

© Kara Richardson.

The board does a lot of work behind the scenes, so it is important that they are recognized as well. It is nice to acknowledge them on stage at fundraisers and provide them with a year-end appreciation gift. Bringing food and wine to the board meetings will also make your station advisors feel motivated and ready to work for the common good!

TIP

Of course, all of the donors who make your station possible should be thanked profusely as well. Handwritten notes are a nice way to let them know how personally grateful you are for their support. Avoid incentives such as hats and T-shirts in direct exchange for donations; I've learned that some donors don't want these rewards and would prefer that their money go to more important purposes within the organization. Likewise, this can add to the confusion at tax time as they would have to subtract the value of the gift from the value of their donation if making a tax deduction (see Chapter 8). Instead, we send our donors special holographic stickers that recognize their dedication.

Funding continuing education

There will always be new things to know, so be sure to keep up with it all. Your station should budget for continuing education opportunities for your board and staff, including conferences and seminars where they can learn advanced skills while connecting with others in the field. If you are new at nonprofit management, there is likely an organization in your community — or one that you can connect with online — that can provide guidance and workshops to help you build a stronger organization.

Conferences are great ways to network, but after travel, accommodation, and registration, they can get quite expensive. If your station can only attend one conference a year, the Grassroots Radio Conference (GRC; `virtualgrc.org`) is where LPFM, college, and community stations meet. This D.I.Y. event is hosted by a different station each year, so you get a behind-the-scenes look at how another organization does it. There are panels, workshops, and social events with like-minded stations that always make the GRC a mind-blowing event. Collegiate stations will also like the annual conference hosted by College Broadcasters Incorporated (`cbi.org`) with a strong radio element that is made just for them. The National Federation of Broadcasters (NFCB; `nfcb.org`) hosts an annual conference, which has a more professional vibe than the D.I.Y. gathering of the GRC. Expect to see many National Public Radio-affiliated (NPR) stations in attendance. This event also boasts an amazing vendor floor where you can preview all the latest audio advancements.

Joining Professional Organizations

You may find it useful to join some professional organizations such as local broadcaster associations or statewide nonprofit networks. Memberships are usually based on station income, and annual fees for young organizations may cost between $50 and $100 per year. You can find out more about suitable associations by searching your state name along with "Association of Broadcasters."

One of the best things about joining with your state Association of Broadcasters is that they often offer a site *mock inspection*, where they will send an engineer to your tower site to confirm that everything is functioning properly and then send a report confirming that to the FCC. For the three years that this pre-inspection is valid, the FCC will trust this information and not show up at your site to do a surprise visit.

The National Federation of Community Broadcasters (NFCB) offers memberships for community stations. They offer support, guidance, and music licensing programs for noncommercial stations. NFCB membership begins at $500 per year.

The Pacifica Affiliates Network (`pacificanetwork.org`) unites over 200 college and community radio stations across the United States. Their mission is "to advance the practice and pursuit of grassroots community media." They offer a vast array of support services including monthly webinars and a range of IT services. Membership provides access to Pacifica's Audioport, a content-sharing network that allows your station to syndicate programming among other affiliates and tap into the vast library of Pacifica programs — including live coverage of international news events and Congressional hearings.

Be aware that the National Association of Broadcasters (NAB) joined NPR in actively opposing the LPFM movement in Congress. Some state broadcasters only offer "non-voting" memberships to LPFM stations.

Budgeting for Communications

Communication is key! Of course, your radio station is the ultimate communicator — but not until people tune in. The beautiful thing about being a LPFM is that folks will be able to stumble across your station by scanning the dial. However, a concentrated multimedia campaign to acquire new listeners will really help things heat up fast! And for that to happen, you need to budget for it.

Email is a free and immediate way to connect with supporters but you need their addresses to get going. Start building your list immediately by seeking out folks in your community who seem important to add. Be sure to always blind copy (BCC) because people will get very mad if you share their email address with everyone else on the list. Websites such as Mailchimp (`mailchimp.com`) or Sendinblue (`sendinblue.com`) offer free email templates and mailing services. These programs make designing an attractive email easy and take the worry out of spilling your address list to the masses.

Social media is, for the most part, free and ubiquitous. Go for it! Make pages and get your DJs involved. You may want to sponsor posts, but ART FM got burned badly by a social media hack that was draining $1,000 out of our account each day for ads in a language we did not understand. Luckily, we stay on top of our finances and never let more than a few days go without checking our balance. We finally got everything cleared up through a maze of chatbots, and let me tell you, that was a terrifying experience. We now have a separate bank account that we use for any kind of online purchases.

One of the biggest communication costs you face is in the old-school print medium. But you should just have fun with that. Make flyers, posters, and postcards. In addition to being sent in the mail, these can be sent or dropped off at local establishments, pinned to community bulletin boards, or handed out at community events. I recommend working printing and mailing expenses into your station budget.

Keeping Financial Records

As a nonprofit organization, you must keep clear and complete financial books. Since the board of directors is fiscally responsible for the station, they must be able to clearly see how the station is generating income and where it is doing its spending.

Bookkeeping

Many organizations hire a business manager, bookkeeper, or accountant to keep track of these expenses and to file the annual taxes. Sometimes this will be the job of the general manager or board treasurer.

Programs helpful for nonprofit accounting include Quickbooks (`quickbooks.intuit.com`), Freshbooks (`freshbooks.com`), and Zoho Books (`zoho.com`). These programs cost $200 to $400 per year. A whole host of accounting and financial *For Dummies* books will help develop your organizational bookkeeping method. For example, *Accounting For Dummies*, 7th Edition, by John A. Tracy and Tage C. Tracy (John Wiley & Sons, Inc., 2022).

Another part of your bookkeeping is tracking listener donations. At ART FM we still use an Excel spreadsheet for this. A simple alphabetical spreadsheet works just fine to get you started. The key is to keep the columns with each year's donations precise. If you prefer, there are much more sophisticated customer relationship management (CRM) systems such as Little Green Light (`littlegreenlight.com`), DonorPerfect (`donorperfect.com`), and Neon (`neonone.com`) that help you record your donor relations. These systems can cost upwards of $1,000 per year but they can generate thank-you notes, renewal reminders, birthday cards, and more. For the D.I.Y. folks, CiviCRM (`civicrm.org`) is an open source, free CRM package; you can sign up for a hosted CiviCRM account or you can download the software and run it on your station's computers.

KEEPING YOUR OWN BOOKS

It is okay to keep your bookkeeping simple so long as you keep everything in clear order. As a nonprofit, your finances will be available to the public and you will be accountable for every dollar. I am still using Microsoft Excel spreadsheets for the ART FM bookkeeping. I create a workbook for each fiscal year, divided into three separate sheets. The first is Income, the second Expenses, and the third Capital Expenditures. Income is pretty self-explanatory. The purchases that go on the Capital Expenditures sheet are different from those that go on the Expenses sheet in that they are items that will last us three years or more. Capital Expenditures would include furniture and broadcast equipment; expenses include things like utility bills, paper products, printing fees, and other disposable goods. This book's online resources include a sample budget template.

I keep a corresponding three-ring binder for each fiscal year. In it, I use heavyweight sheet-protectors that are divided into the three main categories previously discussed: Income, Expenses, and Capital Expenditures. Within each of these categories, I label each of the sheet protectors with the categories I use on the spreadsheet such as Postage, Utilities, Licensing Fees, Station Supplies, Hospitality, and more. When I get a paper receipt, I file it accordingly. Online receipts are preserved in a digital folder.

Trust me, my bookkeeping is simple — but it works!

Filing annual taxes

Remember that every year your Federal 501(c)3 tax-exempt organization will have to file your taxes with the IRS by the 15th day of the fifth month after the end of your organizations' accounting period. That is May 15 for those working to a standard calendar year. Failure to do so in a timely manner will result in expensive fines. Failure to do so for three years will result in the loss of your tax-exempt status.

If your nonprofit makes less than $50,000 in a year, you can simply file the 990-N e-postcard. This asks for little more than a confirmation of your address, list of your board of directors, and a statement of your total annual income. It takes about five minutes to complete. Nonprofits with annual revenue between $50,000 and $200,000 (but less than $500,000) in assets must step up to the abridged Form 990-EZ. The lengthy 990 is required of nonprofits who have a gross annual income of more than $200,000 or assets collectively valued greater than $500,000. All tax filing must be done online.

Maintaining transparency

The IRS states that not-for-profit organizations must make their 990s available for public inspection for a three-year period beginning on the date the return was filed. To maintain the public trust, it is imperative that you keep track of every dollar that your organization makes or spends. Review the IRS website (irs.gov) or speak with your CPA to confirm that you are filing correctly.

If anyone asks to review your taxes, you will be obliged to show them. Posting your 990s on the website Charity Navigator (charitynavigator.org), which provides public access to financial data about nonprofit organizations, helps build your reputation as an open and transparent organization. Investors may be more likely to give if they can see where their money is going.

Remembering Everything Else, Too

No matter how much you plan and try to budget every detail, there will always be something you didn't think about. This could be some small hospitality item or it could be a critical equipment emergency. This section highlights a few odds and ends you may want to consider, but remember to always expect the unexpected!

Saving up for sundries

A number of basic expenses keep your station running smoothly, such as office items and printer ink. For a healthy atmosphere, cleaning products, air purifiers, and hand sanitizer should always be in good supply. Pens, clipboards, and headphone adapters are essential radio gear. Offering complimentary coffee, tea, and hot chocolate to the DJs and guests is a nice touch. This means that you also have to provide cups, stir sticks, cream, sugar, and napkins. If your budget is tight, you can place a donation jar at the coffee station to help offset the costs.

Buying merch

Merch costs money, but it helps to spread the word about your radio station. So budget for it. Every year at WXOX, we purchase fun swag items to give away. Koozies are always popular and 45 adapters are great conversation starters. A personal favorite is the WXOX flashlight pen. We have also given out baseball caps, beanies, and lunchboxes. Go online, search "custom merch," and let yourself get carried away! However, be sure to use local printers and manufacturers whenever possible — your station should be committed to supporting the local economy.

Get lots and lots of stickers. Your DJs and staff members should be armed with stacks of stickers at all times. These are great little handouts and they don't wind up in the trash or on the floor like paper flyers do. Make them cool enough and they will show up on laptops, car bumpers, and cellphones doing all the hard advertising work for you. You especially want stickers with you when you run into celebrities such as rock band Shonen Knife (take a look at Figure 9-2). It's so easy to hand them a sticker, as DJ Clawdada did here, and say, "Hi, can you do a sound drop for our community radio station?" and then record them with the voice recorder on your phone saying, "Hi, we are Shonen Knife and you are listening to WXOX 97.1 FM!"

© Brett Holsclaw.

FIGURE 9-2: Shonen Knife shows off a sticker given to them by DJ Clawdada as he scores a sound drop for ART FM!

T-shirts and hoodies are amazing investments with a good return value. Friends and fans of the station will willingly purchase these items and wear them as walking, talking billboards for your station.

Launching events

A lot of expenses are involved in hosting events but they also provide lots of opportunities for making money (which I discuss in Chapter 8).

Events are excellent community builders and are well worth running. Sometimes you'll attempt to raise money with these events so that they will help build your finances; other times they may just be social gatherings for friends and supporters. If you don't have much to spend, try pot lucks and studio open houses. It doesn't have to be fancy; the main thing is just getting everyone together.

At ART FM, we have been able to make a small investment go a long way with our musical fundraisers. With enough lead time, you can often get food, drinks, and equipment donated in exchange for on-site and on-air recognition.

TIP

To discover more about budgeting for community events, I recommend checking out *Event Management For Dummies* by Laura Capell (John Wiley & Sons, Inc., 2013).

Saving for station updates

Updates and upgrades must happen from time to time, because you want to continue to improve your station's sound and environment. Invest in making your studio warm and welcoming and avoid on-air problems by keeping all the equipment up and running and in good shape.

Make sure that you plan ahead. Keep a rainy-day fund for replacing equipment that breaks or wears out — especially for items such as computers or soundboards that are essential to your broadcast. Keep liquid assets on hand to use for quick fixes and smaller repairs but also start building a secure emergency fund right away.

TIP

Nonprofit advisors recommend having 3 to 6 months of operating expenses squirrelled away in a savings plan that can only be accessed by a board action. This is a noble goal, but in the beginning, there are a lot of expenses and it may take a while until you can safely reach that amount. You will get there!

4

Star Systems: (Low) Power to the People!

IN THIS PART . . .

Understand the legalities of broadcasting and running a Low Power FM station to stay on air and keep out of trouble.

Develop your program content and set the right broadcasting vibe.

Discover the talent in your own community, train them well in broadcast responsibilities, and listen in as they become local stars!

IN THIS CHAPTER

» Remembering to ask for help and legal advice

» Understanding and following federal broadcasting law

» Considering other important legal concerns

Chapter 10

Keeping It Legit: Broadcasting Rules and Responsibilities

Enough with the fun and games! This chapter gets serious. Serious enough that I recommend you treat the entire text as one giant Remember and Warning icon combined.

You have worked hard for your Low Power FM license and you do not want to lose it — or have someone else lose it for you. Unlike online radio stations where pretty much anything goes, FM broadcasters are held to higher community standards. Failure to uphold these standards can result in your license slipping through your hands and all the work you have put into your station disappearing as if it were some desert mirage.

In this chapter you discover how to protect your license by obeying local and federal laws while creating your own guidelines for best station practices. It is the general manager's responsibility to uphold broadcasting law, community expectations, and internal policies (check out Chapter 12 for more on this and other important roles at your radio station). Over the years I've come to realize that the fewer rules I make, the fewer I have to enforce; but as free-form as things remain at WXOX, experience has shown that some rules are necessary to maintain the safety of the staff and the harmony of our station. After you are on air, this will become the most important chapter in this book. Sit up straight, get your elbows off the table, use the outermost fork, because it's time to learn the rules and regulations of American FM broadcasting!

Seeking Help and Getting Legal Advice

WARNING

I discuss broadcasting law in this chapter but remember that I am a station manager and not a broadcast attorney. I share with you what I know but this should not be considered legal advice. When you have legal questions, please consult the Federal Communications Commission website (`www.fcc.gov`) or speak with your attorney.

TIP

You may also contact an FCC Broadcast Information Specialist to help you understand the rules and regulations; reach out to them at (202) 418-2700 or `radioinfo@fcc.gov`. Common Frequency (`commonfrequency.org`) and the Prometheus Radio Project (`prometheuradio.org`) are nonprofit organizations with engineers and lawyers on standby to help LPFM stations when they have compliance questions. Additionally, REC Networks has a reputation for knowing the rules inside and out; their website offers many resources for LPFM broadcasters including a station self-inspection checklist and an online compliance handbook (`recnet.com/lpfm`). Please don't be afraid to seek help or advice.

Understanding FCC Laws

REMEMBER

Let's start at the top. The Federal Communications Commission (FCC) is the federal enforcement agency that upholds the federal broadcasting laws of the United States. If they speak, you'd better listen. If they ask you to do something, you will have to obey or face the consequences. Like you, the FCC is beholden to the laws Congress makes. The laws are not always black and white; they are sometimes shades of gray. There are DJs who like to push the line, especially with those rules that are still to be determined. How far you can push broadcast boundaries is directly related to how much you can afford to pay broadcast lawyers!

Although you may not agree with all federal laws, you are obligated to uphold them. We are very fortunate in the United States that when we feel laws are unjust, we can petition our Congressional representatives in Washington to change them. It is also important to know that when the FCC makes new rules, they open a window for comments from the public. Be sure to act on the issues that are important to you as these comments are read and taken seriously by the commissioners. Find out more about commenting on FCC proceedings at the following website: `https://www.fcc.gov/consumers/guides/how-comment`.

WHAT IS THE FEDERAL COMMUNICATIONS COMMISSION?

Throughout this book, you'll find numerous references to the Federal Communications Commission (FCC). So what exactly is it?

The FCC regulates interstate and international communications by radio, television, wire, satellite and cable in all 50 states, the District of Columbia, and U.S. territories. An independent U.S. government agency overseen by Congress, the commission is the United States' primary authority for communications law, regulation, and technological innovation.

What is the FCC's responsibility?

The limitations on the FCC's power to restrict or ban speech begin with the First Amendment to the U.S. Constitution, which decrees that the federal government "shall make no law . . . abridging the freedom of speech, or of the press." Accordingly, through the Communications Act, Congress declared that nothing in the statute shall be understood or construed to give the FCC the power of censorship over the [broadcast] communications or signals transmitted by any [broadcast] station, and no regulation or condition shall be promulgated or fixed by the FCC which shall interfere with the right of free speech by means of [over-the-air] broadcast communication.

The FCC has long held that public interest is best served by permitting free expression of views. Rather than suppress speech, communications law and policy seeks to encourage responsive counter-speech from others. This ensures that the most diverse and opposing opinions are expressed (even though some views or expressions may be highly offensive).

Nevertheless, what power the FCC has to regulate content varies by electronic platform. Over-the-air broadcasts by local TV and radio stations are subject to certain speech restraints, but speech transmitted by cable or satellite TV systems generally is not. The FCC does not regulate online content.

The FCC does impose certain restraints and obligations on broadcasters. Speech regulations are confined to specific topics, which usually have been identified by Congress through legislation or adopted by the FCC through rule making or proceedings. These topics include

- Indecency
- Obscenity

(continued)

(continued)

- Sponsorship identification
- Conduct of on-air contests
- Hoaxes
- Commercial content in children's TV programming
- Broadcast news distortion
- Accessibility to emergency information on television
- Inappropriate use of the Emergency Alert System warning tones

What are the broadcasters' responsibilities?

Individual radio and television licensees are responsible for selecting all broadcast content that their stations air. This includes both entertainment programming and news/informational programming, whether the broadcaster actually creates the content or acquires it from third-party program producers. Broadcasters also decide how their programs, including call-in shows, are conducted and whether or not to edit or reschedule programs or material (for example, moving a potentially indecent program to a late-night time slot when few or no children may be in the audience).

Delving into The Public and Broadcasting

REMEMBER

The FCC has written a publication entitled *The Public and Broadcasting* (published by the U.S. government in 2008, and revised in 2021) which describes what the public can expect from broadcasters. Although not all the sections apply to your LPFM station, many do. Your station needs to maintain a hard copy of *The Public and Broadcasting* on-site and be able to provide copies to members of your listening audience upon request. You can obtain copies by calling the FCC toll-free at 1-888-225-5322 (1-888-CALL FCC). This document can also be found on the FCC's website: `https://www.fcc.gov/media/radio/public-and-broadcasting`. When you have a question, this publication is a great place to start. This FCC handbook is designed to help members of the public understand somewhat obscure, grayscale laws.

TIP

All on-air DJs are required to know and obey the FCC guidelines within this handbook — no exceptions. All staff members should read and regularly review *The Public and Broadcasting* because a station's ability to acquire and maintain an FM license depends on the ability of their staff to uphold the federal communications law. The station will have no choice but to discipline any staff member that is in violation of these FCC guidelines (see Chapter 12 for more on enforcing your station's rules) or give up the station's license.

Remembering First Amendment rights

The best news within *The Public and Broadcasting* is that the FCC is limited in its powers to regular the content of your on-air broadcast. The First Amendment and the *Communications Act of 1934* are the one-two punch that keep the U.S. government from interfering with your freedom of speech over the airwaves.

The First Amendment states that Congress

> *shall make no law . . . abridging the freedom of speech, or of the press.*

Section 326 of the *Communications Act of 1934* backs that up, stating that nothing in the statute

> *shall be understood or construed to give the Commission the power of censorship over the [broadcast] communications or signals transmitted by any [broadcast] station, and no regulation or condition shall be promulgated or fixed by the Commission which shall interfere with the right of free speech by means of [over-the-air] broadcast communication.*

In light of this, the FCC is still required to uphold certain federal laws that regulate the airwaves. Where the First Amendment ends and the FCC's authority begins is not always clear. The *Communications Act* provides the FCC with the power of enforcement in situations where stations are acting outside of the laws Congress has enacted. This means FCC agents can show up at any moment, seize your equipment, take your license away, issue fines, and even put you in jail — but only if you break the law.

Find official information about freedom of speech over the airwaves in the FCC's online consumer guide: `https://www.fcc.gov/media/radio/public-and-broadcasting`.

Abiding by FCC Laws

The following sections are based on the FCC's publication *The Public and Broadcasting*, an official document that details the laws that govern noncommercial FM broadcasting within the United States. (For more on this document, including a link to the full text, see earlier in this chapter.)

A lot of information in *The Public and Broadcasting* does not pertain to LPFM stations. I've distilled the sections of publication that seem imperative into this chapter, but this should not be seen as a replacement for reviewing the original document. This also should not be considered legal advice. Again, when in doubt, you should always consult a broadcast attorney or the FCC itself. Station managers should read the original FCC text for themselves, to see if there are other portions of the law that may apply to their specific situation.

TIP

The original text of *The Public and Broadcasting* can be found here: `fcc.gov/media/radio/public-and-broadcasting`.

WARNING

The Public and Broadcasting is not optional material. To maintain your FM license, you and all the members of your staff must follow these federal laws.

Abiding by basic laws and policies

Each radio station is expected to abide by the basic laws and policies enacted by the U.S. government. The FCC has the power to enforce these laws, but they must be careful not to interfere with freedom of expression in radio broadcasting. This includes viewpoints that may be objectionable to many listeners. However, the FCC reminds us that "the right to broadcast material is not absolute."

Because the FCC is strictly prohibited from determining what stations play on air, each licensee must use their own discretion to choose what material they broadcast. This applies to all topics including music, news, sports, and public affairs.

The First Amendment's guarantee of freedom of speech protects material that offends and ridicules certain populations. The FCC may not interfere with programming that propagates stereotypes involving gender, race, religion, national background, and more, and recommends that listeners who offended by a station's programming make it known in writing to the station license holder.

REMEMBER

Stations have a right to maintain their own self-determination in regard to programming decisions and have no obligation to host any particular person or group on air. They also have the right to reject any programming that is presented to them.

Avoiding hoaxes and distorted reporting

The FCC does not require or review anyone's qualifications to collect news or report on current events. Decisions of this nature are left up to station management. However, two topics within broadcast news are regulated by the FCC — hoaxes and news distortion are both prohibited under commission rules.

Thanks to the Mercury Theater 1938 production of *War of the Worlds*, which sent listeners nationwide into a panic (hear more about this epic broadcast in Chapter 17), the FCC now prohibits the broadcast of fake crimes or fictional catastrophe if

» The station licensee was aware the information was false.

» The false material directly results in substantial public harm.

» It was foreseeable that such a broadcast could result in harm.

The FCC says that rigging or slanting the news is an act against the public interest. In *The Public and Broadcasting*, the FCC states that it will investigate any complaint against a station where there is evidence of the station rigging or slanting the news and especially where there is direction to employees from station management to falsify the news.

Balancing political broadcasts

Respecting the importance of the free flow of information throughout the electoral process, the Communications Act and the FCC rules have a few specific rules relating to candidates for public office.

If a broadcast station provides airtime to a candidate running for public office then it is legally obligated to provide equal opportunities to all other candidates for that office. This rule applies to candidates for federal, state, or local office for the duration of their campaign.

Certain news programs and public affairs shows regularly broadcast on the station are exempt from providing equal opportunities. Likewise, certain documentaries and on-the-spot coverage of bona fide news events — including debates and political conventions — are not required to provide opposing candidates with equal opportunities.

Watching out for objectionable programming

REMEMBER

The FCC must tiptoe around the First Amendment. They may not make or enforce any laws that interfere with the constitutionally protected right to Freedom of Speech. The FCC can only regulate broadcast content in a few specific situations.

Inciting lawless action

Broadcasters are prohibited from inciting *imminent lawless action*. According to the United States Supreme Court, the government may curtail speech if it is both intended to incite or produce imminent lawless action and likely to incite or produce such action.

Obscene, indecent, and profane programming

WARNING

The FCC maintains rules to protect children against potentially harmful programming and to assist parental control over what type of programming their children are exposed to. Because broadcast transmissions are freely and easily accessible to all — including children — the FCC has shown that station management has the responsibility to protect them.

TIP

The Seven Words You Can't Say on Television came from a comedy routine by George Carlin of that very name and they apply to radio as well. If you are over 18 and not in a public place, I recommend that you listen to this routine — though it is full of profanity — because it illustrates how unclear the lines of "decency" actually are. You can find it online by searching for the "The Seven Words You Can't Say on Television."

Obscene material may not be broadcast at any time. Indecent and profane material may not be broadcast between 6 a.m. and 10 p.m. *The Seven Words You Can't Say on Television* fall within the profane/indecent range and may not be played on air between these times. Some stations keep a list of these words posted in-studio in case there is any confusion. Because we often have children in studio at WXOX, we keep our list in a binder that we share with guests before they go on air.

Material does not have to contain one of these words, however, to be obscene, indecent, or profane. Other words and sounds can convey sexual activity. This type of audio would also be subject to enforcement by the FCC.

TECHNICAL STUFF

For those who need help understanding the concept, I tell new DJs to imagine they are stuck in a station wagon full of grandmothers and preschoolers with no control of the radio. Do not make this journey uncomfortable! It is interesting that the FCC decency rules are based on community standards, which tend to vary by community. For instance, something that is acceptable in New York City may offend the community in Norman, Oklahoma. It is always important to know your audience.

OBSCENE, INDECENT, OR PROFANE?

This information is drawn from the FCC's publication, *The Public and Broadcasting* (`fcc.gov/media/radio/public-and-broadcasting`). It helps distinguish between obscene, indecent, and profane materials. Although the FCC is generally prohibited from regulating broadcast content, the courts have held that the FCC's regulation of obscene and indecent programming is constitutional because of the compelling societal interests in protecting children from potentially harmful programming.

This is what *The Public and Broadcasting* says:

Obscene material is not protected by the First Amendment and cannot be broadcast at any time. To be obscene, the material must have all of the following three characteristics:

1. An average person, applying contemporary community standards, must find that the material, as a whole, appeals to the prurient interest
2. The material must depict or describe, in a patently offensive way, sexual conduct specifically defined by applicable law; and
3. The material, taken as a whole, must lack serious literary, artistic, political, or scientific value.

Indecent material is protected by the First Amendment, so its broadcast cannot constitutionally be prohibited at all times. However, the courts have upheld Congress's prohibition of the broadcast of indecent material during times of the day in which there is a reasonable risk that children may be in the audience, which the Commission has determined to be between the hours of 6 a.m. and 10 p.m. Indecent programming is defined as "language or material that, in context, depicts or describes, in terms patently offensive as measured by contemporary community standards for the broadcast medium, sexual or excretory organs or activities." Broadcasts that fall within this definition and are aired between 6 a.m. and 10 p.m. may be subject to FCC enforcement action.

Profane material also is protected by the First Amendment, so its broadcast cannot be outlawed entirely. The Commission has defined this program matter to include language that is both "so grossly offensive to members of the public who actually hear it as to amount to a nuisance" and is sexual or excretory in nature or derived from such terms. This material may be the subject of possible Commission enforcement action if it is broadcast within the same time period applicable to indecent programming: between 6 a.m. and 10 p.m.

Following other content regulations

In addition to regulating the objectionable programming described in the previous section, the FCC has a few other broadcast regulations that all FM license holders are expected to uphold.

TIP

For a complete and linkable list of LPFM rules, visit the REC Net website (`recnet.com/lpfm-rules`). And as mentioned back at the start of this chapter — so worth reminding you about here as you dig deeper — the FCC retains Broadcast Information Specialists to help you navigate the legal information in its entirety. You can email them at `radioinfo@fcc.gov` or phone them on (202) 418-2700.

Official station identification

All U.S. broadcast stations must identify themselves whenever they sign on or off the air. They must also do so as close as possible to the start of every hour, a natural break in the programming. A good general rule is to make the announcement within three minutes of the hour but you do not have to interrupt a song to do so.

The *official station identification* or *legal ID* includes the station's call letters, followed by the community specified in its license as the station's location. Between these two, the station may insert its frequency, the name of the licensee, and/or its channel number. Most commonly you will hear: "WXOX-LP Louisville" or "WXOX-LP 97.1 FM Louisville."

The FCC has clarified that LPFM stations must include "LP" in their legal ID hourly and when signing on and off. As in "WXOX-LP Louisville." You are not required to use the "LP" suffix at other times during the broadcast day (find out more about the legal ID in Chapter 11).

Station-conducted contests

If a station wishes to conduct an on-air contest, information about the rules and rewards must be fully and accurately disclosed. The contest must then be run in the way it is described. The contest descriptions must be truthful and straightforward, and misleading material involving contest rules of operation including deadlines, selection criteria, and winner selection are subject to enforcement by the FCC.

Broadcast of telephone conversations

The FCC demands that before broadcasting a telephone call — live or recorded — all parties must be aware of the station's intention to broadcast the conversation. However, specific notification is not required when it can be reasonably assumed that all parties knew the conversation would be broadcast. Such cases include when the caller is associated with the station (such as an employee or station contractor) or when they dial up the station during a program that regularly airs call-ins.

Equal opportunities

All radio stations must, by law, provide equal opportunities in employment. Discrimination against a person in relation to race, color, religion, national origin, or sex is prohibited. Religious stations are, however, permitted to require that their personnel meet religious qualifications.

Sponsorship identification

The FCC requires that when money or other consideration is accepted by or promised to the station, the broadcaster must acknowledge sponsorship on air at the time the sponsored program is aired and fully disclose all parties involved. This includes gifts provided to station employees in relation to their on-air activities, news broadcasts, and musical selections (known as payola; see later in this chapter).

Staying noncommercial

A LPFM station is, by law, a *noncommercial station*. This means that LPFM stations cannot broadcast commercials or use the frequency for any personal gain. It seems simple, right? But you wouldn't believe how difficult people can make this.

REMEMBER

Commercials are announcements endorsing products or services and encouraging people to buy them. You cannot do this. According to *The Public and Broadcasting*, however, a noncommercial educational station can thank contributors and those who have provided underwriting donations with announcements naming and describing the contributing party or donor. This means you can — and must — tell your audience who is supporting your broadcast. You can provide factual information about them but you may not use any qualitative language or calls to action that would encourage folks to spend money with this or any station sponsor.

TECHNICAL STUFF

UNDERWRITING ANNOUNCEMENTS ON NONCOMMERCIAL EDUCATIONAL STATIONS

Noncommercial educational stations may acknowledge contributions over the air, but they may not broadcast commercials or otherwise promote the goods and services of for-profit donors or underwriters. These acknowledgments may not interrupt the station's regular programming.

For additional information about the underwriting rules, see `https://www.fcc.gov/eb/broadcast/enhund` and `https://www.fcc.gov/media/radio/nature-of-educational-broadcasting`.

(continued)

(continued)

Acknowledgments may not contain

- comparative or qualitative descriptions
- price information
- calls to action
- inducements to buy, sell, rent, or lease

Acceptable *enhanced underwriting* acknowledgments of for-profit donors or underwriters may include

- logograms and slogans that identify but do not promote
- location information
- value-neutral descriptions of a product line or service
- brand names, trade names, and product service listings

Fundraising in the right way

LPFM stations are commercial-free radio stations, so they depend upon alternative methods of fundraising including grants, memberships, underwriting, and live broadcasting (read Chapter 8 for ideas on how to do this). Although a LPFM station can and should use the FM frequency to fundraise for its own operations, the station may not fundraise on air for any organization other than the license holder itself. That's the FCC rule, and all nonprofits are included. Therefore, you can announce events and activities of other nonprofits but you may not directly encourage donations.

However, in 2017, the FCC enacted a one percent rule, which allows LPFM stations to use one percent of their annual airtime to fundraise for another nonprofit organization. This was done to allow stations to work on behalf of emergency relief organizations at times when it may be critical. Do not abuse the privilege; you never know when you will need it.

Announcing program-related materials

The FCC has determined that a noncommercial station may announce the sale of program-related material by for-profit entities so long as the announcement is in the public interest, not a private economic interest.

The price of the materials must be nominal unless they are offered. More detail can be found at fcc.gov/eb/broadcast/enhund and fcc.gov/media/radio/nature-of-educational-broadcasting.

Steering clear of payola

Payola refers to the practice of accepting money or other considerations in exchange for airplay. It is strictly prohibited in noncommercial broadcasting.

The FCC has rules regarding payments to individuals. Detailed information regarding payola is found in Sections 317 and 508 of the *Communications Act* and the Commission's Rule 73.1212, and online at https://www.fcc.gov/consumers/guides/fccs-payola-rules.

Below is a sample policy ART FM has adopted to conform to those rules:

> Any station employee or volunteer who has any role in the selection of broadcast matter will not, on behalf of their own person,
>
> a) accept money, music, goods, services, tickets, or other valuable consideration from individuals, organizations, associations, or other entities to broadcast a program or program material, or
>
> b) promote any activity or matter in which they have a direct or indirect financial interest

Moving Beyond the FCC: Other Legal Considerations

Of course, FCC laws aren't the only legal considerations that LPFM stations need to keep in mind. A whole host of local, state, and federal laws exist for your non-profit organization to uphold. Make sure that you consult with an attorney and have an understanding of ethical business practices well before you get started. I touch on a few of the key issues in the following sections.

Avoiding defamation of character

Defamation is a statement that injures a third party's reputation. To win a defamation case, a plaintiff must show that a false statement purporting to be fact has been made. In addition, that statement must have been made to a third person, and caused damage or harm to the person or entity who is the subject of the statement.

LPFM stations must be intolerant to false statements that injure a third party's reputation. False or unsubstantiated on-air statements made about businesses or individuals can result in expensive legal problems. Journalistic integrity should be expected and required by on-air hosts at your LPFM radio station.

Understanding music licensing

Music licensing is discussed in detail in Chapter 9, but with a spotlight on legal airplay, you need to keep a few considerations in mind.

The *Digital Millennium Copyright Act* (DMCA) of 1998 places some limitations on music playback but they apply only to online broadcasters. However, most LPFM stations have an online simulcast and therefore must abide by the DMCA guidelines which are posted here: `https://www.copyright.gov/dmca`.

Blanket licenses

With a *blanket* or *statutory license* provided by annual memberships from the Performing Rights Organizations (PROs), a station must not, within each three-hour period, play more than four selections by the same featured artist or from any box compilation, or more than three songs off one album and not more than two of them consecutively.

Direct licenses

You can receive a *direct license* from the copyright that enables you to exceed the limitations provided by the blanket license. Stations interested in getting a direct license should reach out to the copyright holder and ask for permission in writing. The online resources for this book include the direct license request used by ART FM.

Music licensing for archived shows

Blanket licenses cover live broadcasts and archived programs up to two weeks old. A station can post show recordings on its website but must have a system for removing them when the time limit expires. Subscription services such as Spinitron (`www.spinitron.com`) and Pacifica's Confessor (`pacificanetwork.org/pacificainternetpackage`) make this easy. Find out more in Chapter 9.

DJs interested in sharing recordings of their show for longer periods of time are encouraged to contact the licensing agencies individually. They can also work through paid platforms such as Mixcloud (`mixcloud.com`) that have music licensing agreements with licensing agencies already. All DJs should be educated about the law in regards to posting and redistributing archived radio programs. Your station should make it clear that it is not responsible for any sharing or rebroadcasting of archived programs other than those posted on your official website.

In order to redistribute a program in the form of a podcast, you need to get direct written permission from each of the copyright holders individually. This is possible but it can be time consuming to reach them — and expensive when you do. This is why you don't hear too much music in podcasts unless the music is created by the producer or someone they know.

TIP

If you have a program that is more talk-oriented, you may be able to remove the copyrighted material before sharing it as a podcast. Music producers who find that impossible should consider syndicating their programs to other radio stations through the Pacifica Affiliates Network (`pacificanetwork.org`) or PRX (`prx.org`). You can read more about this in Chapter 15.

Using other audio sources

WARNING

Music licensing covers audio that has been distributed in the form of an album or other official release. It does not cover interviews or other intellectual material that may be available online. It especially does not cover news reports or conversations conducted by other news outlets. In short, do not play these kinds of interviews or non-musical tracks on air without obtaining written permission from the copyright holder. Contact the artist or publisher prior to using such work on your broadcast.

A few sources of copyright-free material exist online — including the Free Music Archive (`freemusicarchive.org`) and Freesound (`freesound.org`) — but unless the site specifically states that it is *copyright-free, creative commons, open-source,* or *free for educational use,* you must assume that it is not. More information on locating copyright-free material is in Chapter 16.

Deciding on your house rules

Broadcast law is a relatively new field of legal study and many issues are still to be worked out. If you have enough money for fines and legal fees, you can join the pioneers of LPFM broadcasting and find creative ways to challenge FCC law. You can lobby Congress and push for communications reforms but until anything changes you have to abide by the rules.

Ideally, ART FM would exist in complete anarchy and self-determination. We would not impose any rules beyond those we are required to obey. We have learned, however, the value of establishing our own policies that protect our staff and the longevity of our station. For more advice on developing house rules and setting DJ expectations, see Chapter 12. As things begin to take shape, your station will have to consider what it will take to become a cultural asset to the community as well as a safe and comfortable place to be.

IN THIS CHAPTER

» Planning what to put on air

» Getting ready to go live

» Keeping a watchful eye on timings and guests

» Developing station etiquette

Chapter 11
It's Showtime! Developing Programming and Taking It Live

You made it! There's been a lot of hard work so far, but this is what we all came here for — getting our voices on the FM dial! Radio is such a kind, beautiful medium. So generous, welcoming, and powerful! If you let it, it will support all of your endeavors and amplify all of your causes. Fly with radio to impossible places, get lost in the exotic sounds, then click your heels and wake up surrounded by family, snuggled up in the safety of home.

"I like to do radio shows," Grammy nominee Esther Phillips told Professor Angela Davis in an interview for the Pacifica Network in 1977, "You don't have the robot looking at you. I call the cameras robots, and they are right there on you." It is true that in the absence of an invasive lens, working in radio can be a much more comfortable, private experience than appearing on television — even in front of thousands of potential listeners! On the radio, you are always behind a curtain. You can be whoever you want, dress any particular way, and still be captain of the spaceship, hero of the battlefield, Queen of the Nile, or even the Wizard of Oz. Radio is transformative. I have seen individuals who are painfully shy in person come to life with the most vibrant personalities when they go on air, and vice versa. Not being able to see, feel, or hear your audience is comforting to some, unsettling to others.

The key to great radio is to be creative and expressive, always keeping your audience in mind. Focus on one listener. Think about how they may be feeling and what they may be thinking. Connect with them. Speak *to* them. Speak *for* them. Emotions travel across the airwaves directly into the heart of the listener. Music can save someone's soul, and a smile can be heard through the tears, even from many miles away.

In this chapter, I explore the elements of a successful radio program and provide you with the tools for amplifying local voices on the FM dial. Wherever you want to take it, your radio is ready!

Moving from Dreams to the FM Dial

There are as many approaches to radio as there are people in the world — in fact many more! If only everyone had the opportunity to get on the FM dial, what a planet this would be.

The beautiful thing about community radio stations is that programmers are allowed to be themselves. This shouldn't seem like a big deal but it is, especially in the United States where the FM band is dominated by big corporations who tell their DJs exactly what to say and play while they are on air. To me, it's hard to believe this format could possibly work. On WXOX we have DJs of all musical interests who make the station sound completely different every few hours. Listeners tell us they really enjoy the dynamic programming. They find it a fresh alternative to the ever-predictable commercial stations that play the same thing all day, every day.

Although some LPFM stations offer one specific type of programming, most are freeform, if not by choice, then by necessity. Quite simply, it can be hard to fill a 24-hour programming day with volunteers of one specific genre. It's much easier when you cast with a wider net. Plus, it's great to expose your listeners to a wide variety of music — push their listening experience outside of preconceived categories of what it is they think they like.

Some of the most powerful radio programs are those that are not planned at all — they just happen! They can't help themselves from happening. And you should make room for them to happen. When an urgent event shocks the community it forces us out of our regular programming and into topics we never imagined ourselves covering — such as how to wear a mask during a pandemic or how to stay safe during an active shooting. If a station is locked into a solid programming schedule, it may not be able to break in with live reports, news conferences, and important interviews as they happen.

To me, great radio depends on an open mind and a responsive schedule. The most important difference between radio and on-demand listening such as Spotify, Pandora, and Tidal is the human being on the other side of the line. And how important is that?

TIP

My advice is to keep your station live and let your DJs be expressive. Encourage them to follow their passion and trust their instincts. Be open to their ideas and flexible with their revisions. In today's society, people are given so few opportunities to be creative that when they have the chance they really can come up with the most mind-blowing things!

Music programs

Music shows can focus on specific genres: hip hop, punk rock, classical, jazz, or any other flavor that they can possibly conjure. Focusing on a particular type of music allows the DJs to really build an audience around an interest group. This section includes a few examples of the music programs ART FM broadcasts on WXOX 97.1 FM. *Muddle Instead of Music,* for instance, focuses on contemporary avant-garde composition. There are very few radio programs like this in existence, and host Matt Ertz has developed a following among contemporary composers and musicologists worldwide.

Other music programs are based on a theme, and these shows are likely to include multiple genres. *The Diva Hour* on WXOX with Carly Johnson is a showcase of female singers "ranging in genre from jazz to pop and spanning the decades from the 1920s to the present." Hosts of these radio shows find and follow the common threads that weave across all music and time.

Some programs on WXOX vary their themes from week to week based on topics like the weather, time of year, special events, or approaching holidays. Some people use theme programs to celebrate friends or remember family members, others to mark special occasions, and still more just have a little fun. *The Exploded View* with Creighton Beryl presents a different musical subject each week. Recent themes have included Boomer Tunes, Birthday Bash, Spring Cleaning, Summer Romance, Glam Rock, Weirdo Music, and more. Another program, *Jake's Jumble* hosted by Jake Groves presents a word jumble every week pertaining to the music he is playing. The listeners are encouraged to tune in, wait for clues, and see if they can solve the puzzle.

Some programs are decidedly chaotic, avoiding themes to make room for spontaneous discovery. *Driftless Turmoil* with Brian Manley, for instance, moves from death metal to hip hop to vaporwave to Tibetan opera, all in the course of just one set!

TIP

Personally, my favorite approach to broadcasting is entering the studio empty handed, pulling things off the walls (check out Chapter 7 for more on building a music library). It can be quite a workout! And it's always interesting how one song leads to another, with a story unfolding as you go.

Spoken word programs

Spoken word shows can be divided into news, sports, arts, community forums, interviews, radio dramas, religious programs, and more. Your station may want to focus on a particular type of talk radio, such as social activism, or like ART FM, the arts. Fortunately, at ART FM we consider almost anything art — so long as the subject is approached in new and different ways.

Regardless of the subject matter, a talk program can be brilliant or boring depending on how it is presented. For instance, I rarely watch movies, yet *Film Fatale*, hosted by Mike Elsherif and JohnBen Lacy, is a movie-related program I really look forward to each weekend on WXOX. These guys have such great banter and apply so many contemporary ideas to their conversation that it almost doesn't matter what topic holds it all together.

Over the years ART FM has had the privilege of hosting arts-focused talk programs from local nonprofits such as Louisville Visual Art, 21c, the Asia Institute Crane House, Kentucky Museum of Art & Craft, and more (head to Chapter 14 for more on involving your community). These shows have been wonderful because cultural institutions are always developing new and relevant programming. They have research and talking points ready to go about exhibits and events and it is easy for them to come on air and share this information with radio listeners. Likewise, they have natural interview connections through which they can bring guest speakers and exhibiting artists on air. Working with a variety of cultural institutions has brought engaging content for our station and outstanding publicity to their work.

LPFM radio stations should consider hosting local news or public affairs programs that discuss recent events in the community. Such programs can be enlightening because they invite local members of the community to respond to current happenings. It is important that the hosts of any news programs have some journalistic background or training, which I discuss later in the "Reading the News" section.

WARNING

News reports and interviews from other stations are owned by their producers. And unless they are clearly marked "Creative Commons" or "Copyright Free" you may not use them without permission. In this sense you must be very careful with clips found on online sites that ignore federal copyright law. Additionally, as innocent as it seems, reading stories on air or performing radio dramas without the proper permissions from the author or publisher can also result in copyright infringement. Find out more about using copyrighted material in Chapter 16.

Darrick Wood, host of *Inside a Question,* and his celebrity chef wife Damaris Phillips, have brought a number of radio dramas to the airwaves on WXOX. To do so, they first rearrange the listening area of the studio to make it into a 1930s living room. Friends are invited to take on different roles in their productions of on-air radio plays — but only if they dress of the era. Thematic food and cocktails are provided, and the result is an audio time machine, not only for the participants but also for the listeners tuning in back home. None of this should come as a surprise; Darrick's regular program is an audio experiment in which "each episode focuses on a different set of questions that require your active participation — yes, I'm talking to *you* —to shed light on the answers." During this interactive program listeners have been known to build robots, invent board games, try new foods, text distant friends, and balance spoons on their noses — which makes the perfect segue into the next section!

Experimental radio programs

Artists have had very little opportunity to experiment with the medium of radio. There are many corners of broadcast transmission yet to be explored. What are the essential properties of radio transmission? What does it mean to communicate with so many people at one time? How does this medium affect the listener? What new things can be done with radio that have not yet been done before?

TIP

Great untapped potentials still remain for the radio broadcast medium. The only way we will discover these new possibilities is through experimentation. Every station, in my mind, should provide open airtime to audio artists who are willing to step outside of the DJ box and seek new possibilities for broadcasting (see Chapters 16 and 17).

Wave Farm in upstate New York is a nonprofit organization providing public access to the airwaves for creative use. They offer grants, workshops, residencies, and installation space for individuals and organizations with an interest in transmission arts. WGXC 90.7 FM Arca is the full power station where you can tune in to their experimental broadcasts locally. If you are outside of their listening area, you can pick up a number of dedicated web streams direct from the Wave Farm Transmission Art Installation Park (`wavefarm.org/ta/art-park/installations`).

One of these is Quintron's *Weather Warlock* (`weatherfortheblind.org`), a live audio broadcast created entirely by the elements. The custom-made synthesizer, shown in Figure 11-1, responds musically to outdoor sensors that are controlled by the sun, wind, moisture, and precipitation. The *Weather Warlock* is currently audible through the Wave Farm website: `wavefarm.org/listen`.

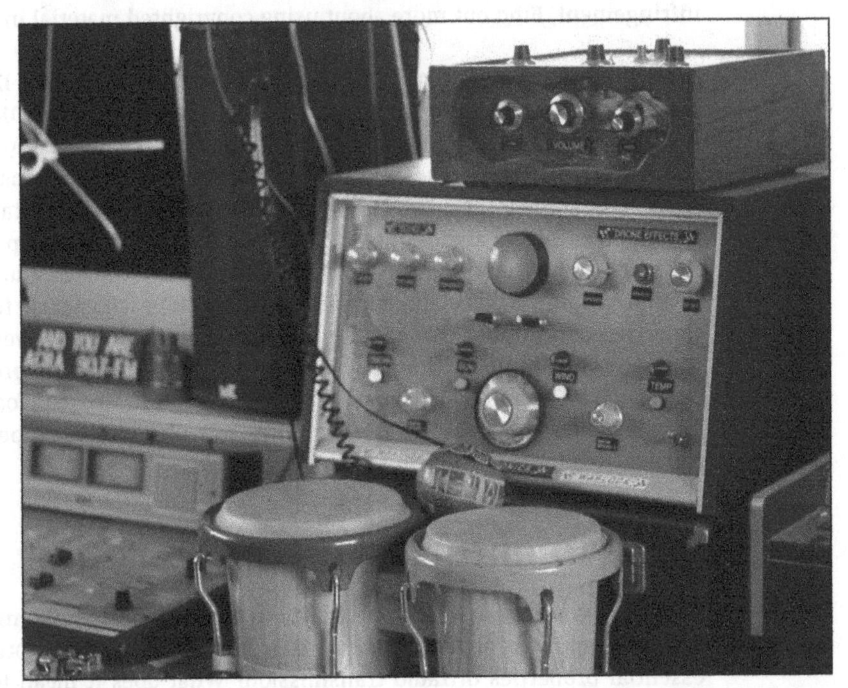

FIGURE 11-1: Quintron's *Weather Warlock* is a weather-controlled synthesizer that works through electronic sensors detecting changes in the weather.

© Jennifer Waits, Co-Founder Radio Survivor.

Setting the Scene

Each radio program, like each radio station, has its own personality. Over the years, it will grow into or out of its original idea. At ART FM we embrace the adventure of broadcasting. We encourage our DJs, likewise, to be flexible with their own radio programs. They should always be seeking new music, discussing new ideas. They should never be forced into a routine, stuck in an algorithm.

Deciding on program titles and DJ names

There is no real formula for coming up with a program title or DJ name. It's a very personal thing. Some folks want to be simple and straightforward; others are quite sophisticated.

REMEMBER

There are a few things to think about when it comes to names. A title should not be locked into a certain timeslot as you may need to move it at some point. For example, WXOX had a show called *Midnight Locker* hosted by Shurtaro Noguchi. It is a great name, but when we moved the show up to 11 p.m., we were in a bit of a quandary — Shutaro came up with a creative solution and changed the name to *Goodnight Locker* and the problem was solved! Other show titles (or program hosts) may not be as adaptive.

Be sure program titles reflect the ethics and character of the station. Recently WXOX had a proposal for a show called *Shotgun!* The concept was that the DJ was going to "shotgun" a lot of random songs — meaning play them in a rapid, random order. However, in today's world where mass shootings are a regular part of the news cycle, this name just did not feel right. The DJ was very understanding when we spoke to him about it and came up with another program name that ended up being a lot more fun. It will be up to your station's discretion as to which show titles you want to add to your schedule. If something doesn't settle with you, let the applicant know; they are usually willing to work with new ideas and if not, perhaps you should heed the signs that this person is not a good team player.

TIP

Catchy program titles on WXOX include *Polaroid City, Rockstar Fever, Stellar Spectrograph, Punk Rockets, Humble Offerings,* and *The Electrostatic Test.* Some folks use their given names to DJ; others will make up something mythological or zany. Mostly, DJ names are super fun, but when people get creative, sometimes things can get a little strange. Try not to judge too quickly and remember you can always talk to the programmer if something does not sit right. As much freedom as you'd like to give to your DJs, remember that you will have to promote their program, so it must be something you can stand behind. Some of the characters you may find on WXOX include Cat Casual, Roxy York, Juan Leone, Lady Jam, DJ Mythos, Sundrop Tom, Anaïs Spin, and Sir Basil Outernational.

Making musical selections

Hosts of music programs spend considerable time picking selections for each week's episode but talk show hosts also will want to consider music that compliments their programs. They will need opening and closing tracks as well as music that can provide an appropriate break in the conversation and give everyone a chance to catch their breath.

TIP

Music can complement any discussion (learn more about developing your in-house music collection in Chapter 7). Try to think carefully about the content and how it relates to your guest and the subject at hand. Additionally, it can be fun to let your guests suggest some tracks for airplay — but make sure that you preview these suggestions to confirm that they are Federal Communications Commission (FCC) friendly.

At WXOX it is interesting to see the different approaches to musical selection that each of the DJs has. As long as they obey FCC laws, we like to give them as much freedom as possible with their musical choices. This is where you will hear their personalities really start to shine!

REMEMBER

Many radio programs have intro and outro songs. This is a great way to define the parameters of your show; now you just have to fill in the gaps! It is a good idea to also have some background or *bed music* to speak over as well. This helps in a number of ways. First and foremost, it keeps the music moving, blends songs together, and keeps talk breaks from being too much of an interruption. Bed music also helps fill any awkward silences that you may have while trying to gather your thoughts on air. If you have a long enough track, you can let it run during a jam, while searching around to find the next track.

Considering guests for your show

Just as I recommend asking all talk programs to think about music that would be a good fit for their show on ART FM (see the previous section), I recommend asking all music programs to think of folks who would be interesting guests on the program. The results have been super impressive! For example, percussionist Tim Barnes was producing a radio special on the music of the Fluxus movement and he was able to get Fluxus artist Alison Knowles to call in for an interview. The result was a mind-blowing program!

REMEMBER

Weekly talk shows can seem challenging in terms of finding new guests for each episode. However, you will quickly find that there are constantly new events and activities that seem important to discuss. Most everyone from artists to politicians will want to come on air and talk about their activities. Businesses, nonprofits, and local production companies always appreciate the free advertising.

Truly in all my years of radio I have rarely, if ever, had an invited guest say no to an on-air interview. I have had guests cancel at the last minute, however, so you always have to be prepared for that — either with a list of local guests that can come in at the drop of a hat or with some other effective replacement.

After your interview programs build a good reputation in your community, you'll find guests coming to you in hope of making an appearance. Artists, musicians, and thinkers will get to know your station as a place for sharing new work, ideas, and concepts. They will start knocking on your door whenever they come into town. It's great when you can create a space where celebrities feel at home hanging out. For more about conducting memorable interviews, see Chapter 14.

Promoting your radio program

DJs should be encouraged to promote their radio programs far and wide and social media is just the perfect tool! Producers can post flashy images with a link to the station website where listeners around the world will be able to tune in.

WARNING

There are dangers of DJs self-promoting, and you may need to enact some policies to make sure all your volunteers are putting your nonprofit organization first. At the very least, you should be prepared to take swift action should something appear on social media —hate speech or otherwise — that can jeopardize the good reputation of your station. See Chapter 10 for more advice.

DJs should be encouraged to make flyers or postcards that they can hand out at shows, post on bulletin boards, and the like. But you may want to have them approved by the station board of directors first (for more on the board, see Chapter 2). You wouldn't believe the things people come up with (check out Chapter 13)! It's cool when DJs start making stickers, magnets, and other swag. T-shirts can be fun but things get tricky when DJs want to sell merch related to their show. It's fair to ask for a cut for the station that makes it happen but you will have to develop consistent policies for all members of the staff. Figure 11-2 shows some of the promotional flyers WXOX DJs have created for their shows.

FIGURE 11-2: Promotional flyers for radio programs heard on WXOX.

RADIO OVER ROBOTS

Shawn Campbell. Founder and General Manager, CHIRP Radio: Chicago Independent Radio Project, shared these views on AI in the radio community:

"I'm sure you know that AI has been in the news lately . . . and radio is not immune. There's actually a company that is pitching an entirely AI-driven radio service to commercial radio owners (listen.streamon.fm/radiogpt). It's quite impressive in its way — it sounds a lot like what most commercial radio sounds like right now — all without the pesky need to pay actual human beings to create it.

This AI service sounds like bad radio. It's faceless, bland and boring. Sure, it can read you a social media post or a news story it's found online, just like actual people voicetracking their shows in another city can Google some things going on in Chicago and drop them in to create a pretend sense of localism, but ultimately, being live (in all senses of the word) in real time is what it's all about when it comes to connection.

I often talk about the extent to which radio has sacrificed its traditional strengths: localism, immediacy, the ability to connect through actual human beings talking to you in real time. These AI services are seeking to create the appearance of all of those strengths. But they cannot make up for the lack of actual human connection.

And as radio and radio-like services become less human, at CHIRP, humanity and connection are what we're all about. It astonishes me that tech companies apparently see the value of the idea of "radio" and are trying to replicate it, while entirely misunderstanding what's at the heart of the whole thing --— real people who listeners can relate to. People who live and work in the community, who can respond to goings-on in that community as they happen, who share listeners' passions, who answer their questions, who are trusted resources who can come to feel like friends.

At CHIRP, we are human-focused radio. We love the music we play. We go out and see bands. We're excited to share our discoveries, to respond to listeners, to talk about what's going on in Chicago. Being on the air is a delight. It's a powerful, meaningful source of connection and community. Every time I'm on the radio, I feel so lucky.

Up with humans! Down with robots!"

Preparing Your Shows

Few things are worse for an on-air DJ than expecting to thrill your listeners with the perfect song and having some technical glitch making it go the exact opposite way. In this section you will discover a few tricks for making sure everything is working smoothly at the start of your show so you can make the right entrance.

Checking equipment

Before going on air, check all of your equipment to be sure it is working properly. There will be a number of DJs coming and going ahead of you and settings may get changed during the course of their shows. Check the volume on your headphones and the pitch control on the record and CD players. Turntable speeds are often a big issue at community stations, with one DJ fuming at the other when they start their 33 on 45! Always put your headphones on before you go on the mic. This is absolutely imperative in terms of making sure you have the levels between your voice and the music even. It makes quite a difference and you wouldn't believe how easy it is to tell when DJs are not wearing their headphones.

Cueing music

Music should be cued up and ready to go before the start of your show. *Cueing* involves getting your song set for airplay — with the needle ready to drop in the perfect position. This can be tricky if there's another DJ in the studio before you, but you will find a way to work it out. Each DJ should be sensitive to the other's sense of personal space.

TIP

Some DJs have theme songs they start with each week while others head right into their playlist. Either way is fine but a theme song can help set the stage for your show, give you time to settle in, and let the audience know that you are back! Add an outro song and some bed music (described earlier in this chapter) and you've already got a skeleton for every show. Bed music is not quite as sexy as it seems — or is it? I guess it depends on your selection. Unless you are doing something experimental, this music should not interfere with the words you are speaking. I recommend instrumental bed without any words or dissonant sounds. You will also want the music to be long enough for you to get in all of your announcements and then some. So either pick a long song or create a sample loop that is set to repeat.

Speaking of dissonant sounds, there may be times when your voice is competing with unwanted background noises in your studio. This can be solved with the use of bed music. The great Clyde Clifford of KAAY made a name for himself in the 1960s for his use of a long atmospheric track by the band Head every time he spoke on air. It sounded really cool but he had a practical purpose — to cover up the sounds emanating from the massive AM transmitter located in the same room!

I recommend keeping all your equipment locked and ready while you are on air. Have music cued up and ready to go in every single audio device possible — records, CDs, MP3s, and even cassettes. That way if one source fails (and it will one day; see Chapter 13) you can rely on your choice of backups.

Going Live!

On-demand listening services such as Spotify and Tidal can provide folks with access to their favorite songs all day long; however, they lack the personality to bring it all together in some kind of significant, meaningful way. The human-to-human connection that happens in radio is something that no algorithm can replace. The knowledge that you are experiencing something real in this world, at this time, in this set of circumstances along with another human being is what really makes radio so special. It is as if the radio DJ is reaching out to you through the airwaves, providing you with guidance, conversation, and a good soundtrack as you navigate your way through the challenges of daily life.

REMEMBER

As a radio DJ your most important job is to be there for your listeners. Although you cannot see them, they are out there, and they are tuned in. Your mood will affect their own and this can have a repercussive effect on the entire community.

Being "in the now"

Always think about your listeners. What kind of conversation do they need right now? What kind of music would set the perfect mood? House music is perfect for a Friday night but not so much on Sunday morning. A talented DJ will be versed in many styles of music. They will be able to adapt to the mood of any situation.

Be Here on WXOX really gets it. Host Emily Miles begins the program each week asking listeners to meditate on the present moment and the power of experiencing this in unison across the airwaves. This immediacy gives radio so much of its connective power.

REMEMBER

When you get on the mic it is important to be in tune with current events and the general atmosphere. Also take note of the programming that has come before your show. If you are coming on air after a 9/11 tribute, you likely don't want to bust right in with a party song. Conversely, if there's a big celebration in your city (Kentucky Derby, Mardi Gras, Tomato Festival — to name but three) you will want to leave your sad songs at home!

From time to time all communities experience some sort of trauma. As much as the DJs may want to provide their listeners with an escape from harsh realities, it is callous to carry on business as usual in certain circumstances. It is important to honestly address the reality of the situation and through this find ways to truly uplift a struggling community. This may not be fun, but it is important.

Holidays, conversely, are a really fun time to be on air, and DJs always like to celebrate! There are a few radio traditions. Of course, every New Year's Day begins with a rendition of "Auld Lang Syne" as champagne bottles pop at the stroke of midnight. On Thanksgiving it is common for American stations to play the 27-minute antiwar song "Alice's Restaurant" by Arlo Guthrie. On October 30, stations replay the Orson Welles rendition of *War of the Worlds* (see Chapter 17). Follow that with Halloween, which is a favorite holiday among American DJs; Rock n' Roll is full of creepy songs, and with spooky effects, you can easily turn the studio into a haunted house or a mad scientist's laboratory!

Setting the scene

With radio you can transport your listeners through outer space to a planet of your own creation. Or you can travel back through the ages to walk among an ancient civilization. To get there you don't need fancy costumes, expensive sets, or the latest special effects. All you need is the power of sound.

In one of my favorite episodes of *Hundred Proof Fountain*, Captain Scratch and Madame Greeneyes transformed the WXOX studio into a slow river in a lush landscape. As they paddled their way down the audio stream, they heard a distant guitar growing louder and louder. Alas, when they found themselves just under it, there was their friend Alexander Sings, serenading them from the trees. I mean, what an entrance for an in-studio guest!

Handling surprise situations

Radio is full of surprises. You just never know what is going to happen and that is what makes it so exciting (for examples, see Chapter 13). It is also why I find podcasting such a bore. It's like watching hockey after the game is over. The thrill is gone. You know how it's going to turn out.

On live radio you have to be ready for anything and you can't be afraid to wing it. Mistakes will happen and technical errors will occur; you've got to just roll with the punches. When a glitch happens — no matter how severe — you have to move on and let it flow downstream. You can't undo the past but you also don't need to step back into it. That is another beautiful thing about FM radio. It happens and then it is gone. It vanishes into thin air!

Keeping Time with the Programming Elements

If you want to know the difference between a community radio station and a public radio station, the key is in the clock. At National Public Radio (NPR), the clock rules. This means that the broadcast day is scripted down to the second regarding what is going to play, what the host is going to say, and so on. There is little room for improvisation. At community radio stations, conversely, it is all about the spontaneity. There are just a few scripted moments during the broadcast hour and the rest is left up to the DJ. Since you never know what is about to happen, you must keep listening all the time! This radio is exciting!

REMEMBER

Even in these free-form environments there are announcement milestones that DJs are required to say on-air. This will vary from station to station but will, at the bare minimum, require DJs to announce the *legal ID* (call letters and city of license, detailed in the next section) at the top of every hour. Many college and community stations have adopted what is called a *peace clock* to segment their programming hours. This schedule is called the peace clock on account of the way it looks when it is drawn out, but I'd say it is more than a coincidence. When people have the opportunity to express themselves freely, as they do with this type of schedule, a peaceful community is the result.

As shown in Figure 11-3, the legal ID must be announced at the top of the hour. The host may opt to read some news headlines as well. At the :20, and again at the :40, the DJ reads a public service announcement (PSA) and a station promo. A local event calendar is read at the half-hour.

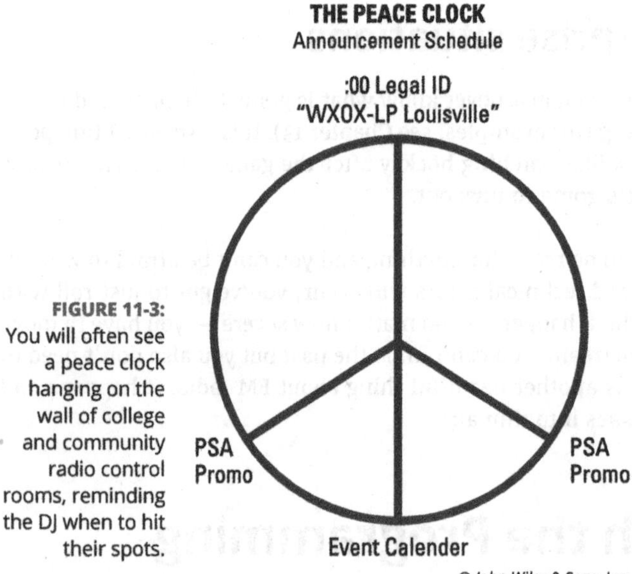

FIGURE 11-3: You will often see a peace clock hanging on the wall of college and community radio control rooms, reminding the DJ when to hit their spots.

Announcing legal identifiers

Due to protections provided by the First Amendment and the *Communications Act*, the FCC is not permitted to tell stations what to say on air — with the exception of one thing, the legal identifier (shortened to *legal ID*). Every station has a legal ID assigned by the FCC. This is your call letters and your city of license. For example: "WXOX-LP Louisville." For more guidance on what you can and can't say on air, head to Chapter 10.

The FCC requires that you announce your legal ID at the top of every hour (:00) or at the closest reasonable break. At WXOX, we instruct our DJs to get it done within three minutes of the hour. It's amazing how many ways people can mess that up. There are just a couple of things you can add to the phrase, such as the frequency number, but I recommend keeping it simple. The FCC states that low power stations must announce the "LP" suffix with their call letters as part of the hourly legal ID but do not have to use it at other times.

Reading the news

Local newspapers are disappearing at alarming rates across the country. Newsrooms have been shuttered and cities as big as New Orleans and Pittsburgh no longer have daily papers in print. As readers move online, advertising dollars move with them and out of the local community. When the papers can no longer afford to pay their journalists, local stories get lost, and local issues get ignored.

This is where LPFM stations have an opportunity to step up and fill in the gap! Developing a first-class newsroom can build a strong reputation for your organization, especially since they can tackle issues that bigger stations may skirt fearing cuts from corporate sponsors.

It is important, however, that your volunteer news team be educated on journalistic principles prior to getting on air. There are lots of ways the novice reporter can get in trouble and open your station up to lawsuits. Defamation of character can be a particular problem for those who haven't been properly trained. If volunteers do not have a background in journalism, look to enroll them in citizen journalist courses locally at community colleges or online through sources such as the Poynter Institute or the Pacifica Affiliates. Michigan State University offers a free online program through Coursera (`coursera.org/specializations/become-a-journalist`) that teaches the skills and ethical standards necessary for gathering and reporting news.

All DJs at community radio stations should be ready to provide important news updates when necessary. Staff members should stay on top of current events and incorporate them into their programs whenever possible. All on-air DJs must be familiar with federal and civil law regarding news broadcasting. No hoaxes, no fake news. See Chapter 10 to review the FCC rules.

The following sections tackle some key areas of news reporting to consider.

Breaking events

The primary job of community broadcasters is to keep our audiences educated and up to date, especially when it comes to important safety information. For example, there was an active-shooter situation at a nearby institution. The WXOX on-air DJ Ricky Erratic was able to warn listeners to stay away from the area, or stay indoors if they were already there. Weather and other emergency alerts will happen automatically thanks to your EAS system (see Chapter 4) but if they feel comfortable, the live DJ may provide updates as they happen.

TECHNICAL STUFF

Civilian coverage of everyday events is critical to our democracy. Technology in the hands of the people has allowed us to witness the truth of situations that the authorities may prefer to sweep under the table. WXOX felt our importance in this regard during the Breonna Taylor protests that occurred in our home base in Louisville. The footage the local TV stations aired had come to them directly from the Louisville Metro Police Department. We, however, were airing livestreams from within the protesters' marches and encampments. Our listeners were getting a different story.

Public affairs

The opportunity for free, informed debate is an important service that LPFM radio stations provide. Public affairs programmers in particular have a responsibility to facilitate the fullest possible discussion on the issues at hand through in-depth questioning of guests representing multiple points of view. The presentation of a range of perspectives can be considered central to your mission.

WARNING

Both programmers and guests should be able to identify the source(s) of any news information they present on the air. Listener participation should be encouraged when possible. Expressions of opinion by programmers and station listeners should be clearly be framed in disclaimers such as "This program expresses the beliefs of the programmer and does not necessarily represent the beliefs of this station." These announcements should be made before and after controversial statements and programs.

Equal airtime

REMEMBER

The FCC states that all noncommercial stations must provide equal airtime to politicians running for public office. This means that if you invite one candidate on air, you must offer all of their competitors' airtime that is equal in length and listenership. This applies also to bills that are coming up for vote. Both sides must be equally represented. See Chapter 10 for more details.

LPFM stations are required to keep a file documenting their conversations with political candidates.

Making public service announcements

In its role as a service to the community, college and community radio stations commonly read public service announcements (PSAs) twice an hour, at :20 and :40 after the hour (see Figure 11-3).

TECHNICAL STUFF

There is no law that requires these announcements be read but it is established precedent. The tradition arose from the early college broadcasters who wanted to be sure that they had clear examples of how they were serving the community with their broadcast should the FCC come knocking.

A number of resources exist for prewritten and prerecorded PSAs. The Ad Council, for example, has worked with the American Red Cross for over 50 years in the production of PSAs of community importance. Over the years, these PSAs have included everything from fire prevention to bouncy house safety. You can access their treasury of written and prerecorded PSAs by visiting `https://www.adcouncil.org/all-campaigns`.

Stations and DJs should not shy away from creating their own PSAs based on current events and local issues. This can be done by reaching out to nonprofit organizations you admire and letting them know about the opportunity.

TIP

You may want to use Google Forms, Cognito, or another online service to create a PSA submission page where local nonprofits can submit their own announcements. You can preview the ART FM example at `artxfm.com/psa`. (Remember, when you accept PSAs from other nonprofits you cannot use the airwaves to fundraise for any organization but your own, the LPFM license holder. This includes promoting ticket sales or incentive programs. For more on FCC law, see Chapter 10.)

It is okay to make service announcements fun. You want them to get your listener's attention. Someone dropped a backbeat on a PSA about Hepatitis B at my college station WRVU 91.1 FM, and it turned out so good that it was one of the most played tracks in the house!

Here are some example PSAs ready for your rotation:

- **PSA Audubon Society:** "The National Audubon Society protects birds and the places they need — today and tomorrow, throughout the Americas — using science, advocacy, education, and on-the-ground conservation. Get involved at www.audubon.org."
- **PSA: Contact your U.S. Representatives:** "U.S. citizens wishing to contact their Representatives in Congress about legislative issues can visit www.house.gov. Click the "Representatives" link for a state-by-state directory that provides weblinks and current contact information."
- **PSA Kitchen fires:** "As you get busy in the kitchen this holiday season, this station reminds you that grease fires cannot be extinguished with water. If a fire starts on your stovetop, turn off the burner and use a metal pan or cookie sheet to cut off oxygen and smother the fire instead."

TIP

Although you want locally-based announcements for nonprofits in your area, this book's online resources include a starter list of PSAs. The Virtual GRC operates a National PSA exchange on their website (virtualgrc.org/psa).

Traditionally, college and community stations have used boxes of notecards in the control room for their PSA and promo scripts (see the next section for more on promos). However, most stations went paperless in 2020 and now it is common to post these PSAs and promos in an online document-sharing system such as Google Drive (google.com) or Dropbox (dropbox.com) that can be bookmarked on the station computer and accessed by station staff in studio and at home.

Reading out promos

Promos are a way that stations can use the airwaves to promote themselves, their programs, and their upcoming events. All programs need their own promo that can be read on air and posted to the station's website. These descriptions should be friendly invitations for tuning in to a particular show or event. Two to three descriptive sentences will usually suffice.

TIP

Here is the magic formula: The shorter the promo = the more often it will get read on air. Very creative promos are the exception to the rule!

Some of the fun promos on WXOX read as follows:

- ***Bases Loaded Radio*** **with Antonio Carlos Echevarría:** "Tune in at the tail end of the week for hit after hit from a wide selection of R&B, hip hop, jazz, and indie music. Host Antonio Carlos Echevarría keeps the line moving with a mix of the latest singles, out of left field deep cuts, and everything in between. Sundays at 8 p.m."

- ***Arcade Automatic* with DJ Maully:** "The beeps and bumps for your leaps and jumps. Enjoy the perfect soundtrack for the game of life. Have fun automatically! Tuesdays 7 to 8 p.m."
- ***Party Pants* with Jaxon Lee Swain:** "This just in — the weekend starts on Wednesday! Take it sleazy with Jaxon & Co. as we fix up for a weeknight rage with a soundtrack of rock & roll, punk, garage, and glam from around the world. Party Pants. Wednesdays 10 p.m. to midnight."

Detailing an events calendar

If you are lucky, you will find someone on your staff who is willing to keep a weekly events calendar. This falls under the promotions director's purview (see Chapter 12).

An *events calendar* can be a brief listing of some interesting upcoming shows. Even two or three events each day will be a service to your community, and folks will tune in to hear it, especially if it happens on a regular schedule. Using online document sharing can help keep it up to date. Abiders of the peace clock detailed earlier in this chapter will read the events calendar at the half-hour (:30).

Keep all announcements FCC legit by sticking to just the facts: Where, When, Who, Why, and How. No endorsements. No calls to action. (See Chapter 10 for more details.)

Hosting On-Air Guests

At ART FM the on-air DJ is responsible for everything that happens on air during their shift. They must be in control at all times, ready to turn off the mic if the guest gets carried away or forgets their FCC rules (as outlined in Chapter 10).

It is imperative that all guests know what they *may* and *may not* say on the airwaves. Many of them, in the age of podcasting, may not realize that we are playing by FCC rules. Embarrassing though it may be, share with them the list of "Seven Dirty Words" prior to going on air, also discussed in Chapter 10. Remind them that other graphic material is off-limits as well.

TIP

Create a brief info sheet to send to your on-air guests well ahead of their scheduled interview or performance. Among other things, make sure it contains directions to the studio as well as a brief list of broadcasting do's and don'ts. The online resources for this book include a sample guest info sheet. You can customize this for use at your station.

Unfortunately, at ART FM we've had a couple of guest incidents that ruined our open-door policy for all the rest. Now we have an online guest permission form that must be completed by the DJs one week prior to the on-air appearance. It is just a few questions and the purpose is to prevent any troublesome guests from getting on the airwaves and putting our license in jeopardy.

REMEMBER

For liability purposes, it is important to have guests sign a performer release form. Get an attorney to help you draft this initial document so that you can have the rights to broadcast and redistribute the performance as you see fit now and in the future. You may also want to include the right to any images taken of them during the on-air performance. This document should also remind them that the broadcast is subject to FCC rules and that they should act accordingly.

You can have your documents available for signature online using programs such as Adobe Sign (`adobe.com`) or DocuSign (`docusign.com`). You may also have your performers sign physical copies; keep a number of these forms printed out and available in studio. However, it is awkward to spring this sort of thing on a guest right before they go on the airwaves. Therefore, it is best to send them a copy of the form via email so that they can review it before arrival.

TIP

Be sure to remind guests to post a link to the station's website on their social media sites so that their fans will know how to tune in. One tweet from a recognized artist can increase your listenership exponentially!

Minding Studio Etiquette

After the scene is properly set, the station will become a sacred space. It will be a home away from home for your DJs, a club house, a temple, and a meeting room. It will be a space where people feel creative, relaxed, and most of all safe. All DJs should do their part to keep the station clean, comfortable, and well maintained. This includes alerting station management if there are any problems at the station — with the facilities, personnel, or the equipment — so that they can be corrected as soon as possible.

In the world of radio, as in anything else, it is very important to keep a harmonious relationship with your neighbors. Minding studio etiquette helps you maintain a healthy relationship with those who share the space with you.

TIP

A smooth transition between DJs is imperative but it can be tricky to achieve — especially in a free-form station when neighboring shows may be very different from one another. Here are a few tips we suggest at ART FM to help make the changeover go smoothly:

- **Arrival**
  - Be prepared to start on time.
  - Arrive at the station at least 15 minutes early, though don't enter the DJ booth more than 5 minutes early.
  - Be courteous to the on-air DJ. Do not disturb the on-air program.
  - Review previous DJ logs to confirm that you are not repeating the same songs, artists, or announcements.
  - Has the legal ID been read or do you need to do it?
  - Pay attention to the vibe of the outgoing show. Transition smoothly out of that mood into your own.
- **Departure**
  - Towards the end of your show, begin to tidy up the control room.
  - Start making announcements about the upcoming program.
  - Encourage folks to stay tuned in.
  - Be courteous to the next DJ. Do not extend your program into their airtime.
  - Give incoming DJ space and time to prepare for the transition — let them connect their computer, load turntables, select promos, arrange mics, and so on.
  - When your program is over, clear out of the control room as quickly as possible to allow incoming DJ space to work.
  - Clear studio of guests and other distractions that may disturb the incoming DJ.
  - Do not overstay your welcome!

ON-AIR EXPECTATIONS: A CHECKLIST

- Legal ID must be announced at the top of the hour. This is FCC law!
- No unauthorized broadcasts of the emergency alert tones.
- No obscene language or descriptions, graphic violence, or sexually explicit material.
- No promotion of commercial endeavors. No prices — ever!
- No defamation of people or businesses.
- No fundraising (except for your station).
- No qualitative words regarding commercial products or events, such as "great," "amazing," or "the best."
- No calls to action, such as "check it out," "go to this show," or "buy this record."
- All political parties and candidates must receive equal airtime.
- All on-air guests must sign release forms.
- All on-air calls must have permission of all parties.
- Say the station name as often as possible.
- Announce the song titles and artist every three to five songs.
- Never say the words "PSA" or "promo" on air.
- Never "um" or "ah" on the air.
- Adhere to the *Digital Millennium Copyright Act* (DMCA) guidelines:
- No more than four selections by the same featured artist or from any box compilation.
- No more than three songs off one "phonorecord" — no more than two of them consecutively.
- Watch your levels. Keep them uniform and out of the red.
- Always wear headphones when on the mic.

IN THIS CHAPTER

» **Putting together your backstage crew**

» **Training your staff and volunteers**

» **Making sure your staff know their responsibilities**

Chapter 12

Forming the Band: Building Your Crew

Of all the things I write about in this book, nothing, simply nothing, is more important than the crew of individuals that are steering your ship. A few motivated volunteers can navigate through the stormiest of seas but one Jonah can sink the entire ship. I have learned, the hardest ways, that although your station must be open to the community, it must be protective of itself.

As you recruit new members, as important as it is to seek talent in hidden places, it is equally important to get to know your volunteers before letting them in too deep. You see, an FM frequency is a powerful thing. It is a great responsibility and you must be able to hold on to it without letting it drive you mad. If this is sounding a little like a passage from the *Lord of the Rings*, I assure you, it is! Your life as the operator of a LPFM radio station will be full of adventures you never imagined from your comfortable armchair in The Shire. Radio, even community radio, is the craziest business I have ever seen. Don't let this scare you, but buckle up for a wild ride!

Everything depends on fortifying your station with a strong, honest, and hard-working crew. Radio is a 24/7 operation and no one person can do it alone — at least not for long. Your radio station not only serves the local community but it builds its own culture of DJs, engineers, fundraisers, board members, and more. Captain of the ship, it is the station manager's job to recognize talent among the staff and employ it in the most effective ways.

In this chapter I discuss tips for recruiting team members and assigning roles and responsibilities to help keep your crew in shipshape and Bristol fashion!

Developing the Station CORE

The DJ's responsibilities to the station do not end when their program is over. In fact, as soon as one show is complete, they have to begin preparing and thinking about the next. Many DJs spend exponentially more time planning their shows than they do implementing them. You will be astounded at how much they will give to producing their radio programs each week.

To survive, however, a station needs more than stars. It needs librarians, bookkeepers, house cleaners — and it needs technicians, designers, and cheerleaders. It needs plenty of brain power, backed up with some muscle. Don't worry at first if you don't have all of these roles filled. You may have to start by wearing a lot of hats but eventually you will see your staff members step up and help share the burden.

REMEMBER

At the heart of any station is a crew that's working overtime to make things work. Some stations call this group the *E-Staff* (short for Executive Staff) but this always seemed too stiff and hierarchal for our community radio environment at ART FM. Instead, we call station leadership simply the *CORE* (short for Critical Operators and Radio Enthusiasts). Station volunteers are welcome in and out of the WXOX CORE depending on their availability and the station's needs.

The station should have at least one or two paid staff members that are committed to moving the operation forward and handling all the critical business (more on this in Chapter 9). Many other positions can be volunteer driven, at least until the station develops a healthy budget for salaries.

TIP

Always be sure to recognize those who work hard behind the scenes and provide ways for others to recognize them as well. For instance, ART FM outfits our CORE staff with special lanyards. These lanyards help them stand out as station leaders at our fundraising and remote broadcasting events. We recognize the CORE team members and thank them for their hard work online, on stage, and in special publications.

The following job positions are traditional at college and community radio stations and may be included in your station's CORE:

- **General manager:** Coordinates all actions of the station. Delegates responsibilities, makes decisions, and maintains day-to-day operations. All other positions report to the general manager who, in turn, reports to the board of directors.

- **Chief operator:** Maintains the technical operation of the station in conjunction with consulting engineers.
- **IT specialist:** Maintains computer systems necessary to the operation of the FM broadcast and online stream.
- **Web designer:** Builds and develops the station's website.
- **Program director:** Works in conjunction with the general manager to schedule daily programming. Maintains promo announcements and develops a printable program guide.
- **Promotions director:** Produces and distributes promotional material, maintains social media and press relations, and coordinates ticket giveaways.
- **Development director:** Manages fundraising campaigns and member relations, and underwrites programs, partnerships, and grant applications.
- **Business manager:** Manages correspondence, payroll, billing, and licensing. Maintains office supplies and other essentials.
- **Music director:** Develops the music library, label relations, and new music acquisitions. Organizes ASCAP, BMI, and SESAC reporting (see Chapter 9 for more on these reports).
- **Local music director:** Coordinates local music rotation and directs the local music show. (The station may also have other types of genre-specific music directors.)
- **Training director:** Manages the training of new DJs and staff members. Schedules shadowing sessions and administers training exams.
- **News director:** Oversees all news programming and newscasters.
- **PSA director:** Manages the rotation of live and recorded public service announcements (PSAs).
- **Graphic designer:** Works with promotions director and web designer to develop visual campaigns for station activities.
- **Studio manager:** Maintains studio and supplies. Manages guest appearances, station paperwork, and Federal Communications Commission (FCC) files.
- **Station archivist:** Maintains and updates physical and online archives.
- **Station interns:** Work alongside members of the CORE staff to fulfill general station duties.

In a station's early days, the founders may play many if not all of these roles but you should have a long-term plan for delegating. This is important for your sanity and to the long-term existence of the station. Think about how these positions will be filled and if they have term limits. Some college stations, for instance, have one-year terms with outgoing officers selecting the individual that takes their place. Community stations may use more of a formal application and election process in their long-term placement decisions.

General managers must be adept at plugging volunteers into tasks that suit them, providing them with tasks that they enjoy, and maintaining the enthusiasm that they bring to the organization. Sometimes their talents and interests do not fit under an established job role. It is okay to listen to their ideas and create custom titles based on the work they are interested in doing rather than forcing them into predetermined roles. For instance, event specialist, support specialist, and director of engagement are all job titles WXOX volunteers have made up for themselves because their work does not necessarily belong inside one specific category. I recommend the flexibility of this approach. After all, this is your station. You make the rules!

Recruiting and Training Your On-Air Team

There are some pretty serious considerations to think about before letting folks go on air. There are plenty of ways for folks to mess up if they are not prepared. We have a vigorous training program at ART FM that prepares our staff for careers in professional broadcasting. A number of volunteers, most of whom had never had broadcast experience before, have gone on to become on-air personalities with NPR, Clear Channel, and IHeartRadio stations. One of our DJs (Jailen Leavell) has recently become a weekend anchor on TV station WDRB (and yes, I'm dropping his name because there's no telling where he is going from here!).

Although it is hard to lose your best DJs when they move on to bigger platforms, this is a time to congratulate yourself and recognize how much your organization is elevating the community. In addition to broadcasting experience, station volunteers gain production skills, organizational approaches, and public speaking opportunities that can advance their careers in a wide variety of professions. And guess what — there is always more talent in the wings!

This section looks into the process for finding stars and preparing them for a brilliant future.

Thinking about the schedule

Scheduling can often be one of the most difficult parts of running a station. Programming decisions can be tough but they determine the culture of your community. One thing to keep in mind when scheduling is that your primary obligation is to the thousands of people in your listening audience, not to any specific DJ applicant or any other single individual.

Don't bring a DJ onboard if you don't think you will have a place for them on your schedule. However, there will likely be plenty of fill-in opportunities for those willing to wait on standby. At ART FM we've found that asking prospective DJs to provide us with their top three timeslots and a list of times they are not available helps us make good placement decisions. We try for one of their top choices but often people just have to settle for times that are available.

Stations have different approaches to scheduling. At ART FM, some of our shows have been in the same timeslot since we began. *Party Pants, Parallel Universe, Ear to Ear, At the Helm, Mythic Beat,* and *Flanger Magazine* are so entrenched in our schedule that they have become a part of our local culture. They have built audiences who count on hearing these DJs at certain times each week and there may be a city-wide revolt if we moved any of these programs now! Freeform Portland KFFP-LP (`freeformportland.org`), however, keeps their schedule fresh by having their DJs reapply for shows every six months, constantly mixing things up! Like many stations, Freeform Portland uses a giant workboard to move program ideas around as they brainstorm the schedule line-up.

Handling applications

When you are ready to bring volunteers onto your staff you will need to develop an application process. ART FM has a program application available on our website (`artxfm.com/participate`). Online software such as Cognito (`cognitoforms.com`) or Google Docs will allow you to build an online form that assembles an applicant's answers in an Excel spreadsheet. These are very intuitive and easy to use. Cognito has the advantage of accepting payments and secure signatures if you need them. Google Docs has the advantage of being free!

Develop a *programming committee* with the purpose of reviewing the submitted applications, seeking shows that uphold the station mission and serve a need within the community.

Remember that you cannot sell airtime on your noncommercial station. Therefore, you should never charge a DJ a fee for hosting a timeslot. You can, however, charge fees for the educational classes that you offer your volunteers — including workshops and DJ training courses.

Vetting new recruits

The most important job of the programming committee is to vet prospective DJs. Carrying out name searches online and in social media sites often provides some insight into the character of the applicants. I recommend asking for at least two professional references as well. At ART FM we do not always pay for an official background check but we do always spend some time online reviewing news articles or public media postings.

For the safety of your staff, carry out these checks before a new person joins your team so you don't find out too late that they are into some kind of violent or hateful material. Sometimes a quick search can let you know right away if an individual seems toxic to your station environment or even dangerous to your station staff. Conversely, you may find some really inspiring information about the individual that encourages you to recruit them for the team and spotlight some of their extracurricular activities on air.

After the programming committee has found the talent they want to bring onboard, it is time to begin the orientation process. Hands down, there is no substitute for getting to know someone in person. Your first step, after vetting the new applicants online, should be to invite them to the station for an in-person meeting.

Overseeing orientation

Because it would take so much precious time to interview each applicant individually, I recommend an orientation session where you can meet all the prospective DJs at once. Give them a tour of the station, and get a sense of who they are and what they may bring to your rotation. These can be informal meetings of any size, but somewhere between 5 and 15 individuals makes a comfortable group.

This orientation is a good opportunity to provide new recruits with a history of the LPFM movement and a history of your station. After that, it is nice to allow them some time to describe their program proposal and any radio experience they may have. I find it interesting to hear their radio concepts firsthand and the other applicants always seem to enjoy hearing one another's ideas.

Putting your recruits through basic training

After you have gathered a batch of new recruits that seem like they will be interesting, responsible program hosts, the next step is to brief them on the rules of FM broadcasting and any regulations of the house.

TIP

Make your radio training a conversation rather than a lecture. You'll find most people are interested in discussing the ins and outs of broadcasting.

Rules and regulations training

WARNING

You may want to dive right into FCC laws, house rules, and other broadcasting regulations (see Chapter 10 and make sure they read it, too!) during an initial orientation meeting or you may want to save it for a follow-up session. Either way, it is important to do this fairly early in the onboarding process so folks can decide if they are up for it or if they want to jump ship. It's better for everyone to find out now if they are not able to uphold the expectations rather than waiting until they've logged many hours of the station training.

Equipment training

After teaching your newcomers the rules, show them the ropes! If possible, find an hour on the schedule in which you can get the entire group in-studio and show them how all of the equipment works. Invite them to bring sample music to play and allow them an opportunity to get on the mic and push some of the buttons. Nothing better than throwing them live on air and seeing if they can swim!

Shadowing

After they understand how the equipment works, partner the new recruits with a seasoned DJ who plays the same music format, so they can observe how the whole program comes together. This is called *shadowing.*

Arranging a shadow session gives the new recruits the opportunity to get a different perspective on program development and to ask any questions they may have been hesitant to ask in the group setting. The host DJ should provide them with some opportunity to get on the mic by reading a station announcement or two and work with the equipment by cueing up a few songs.

TIP

Let your trainees shadow as many times as they like until they feel comfortable going on air. Usually a session or two will do it.

Reverse shadowing

When they are ready, you can allow the new recruits a *reverse shadow* session in which they host a teaser radio program live on the air. Allow your new recruits a one-hour broadcast to strut their stuff. During this broadcast the new recruit should be joined in studio by an experienced DJ who can help out should they need it (make sure it is someone who will be a helpful guide). The new recruit should be encouraged to relax and have fun, rolling with all the punches.

Subbing

If your new DJ has a successful reverse-shadow session they can then start picking up substitute shifts (or sub-shifts) whenever you feel they are ready. The programming committee should listen and hear how the program sounds on air. Sometimes shows turn out very different in reality than they appeared on paper — for better and for worse. If possible, try the new program in a few different timeslots before giving it a home on the schedule.

Taking flight

The FCC no longer requires a test or an operator's license for individual DJs. They now leave it up to the station to make sure that their programmers will not violate federal laws or otherwise jeopardize the FM license. Therefore, it is totally acceptable for a station to instigate its own test for operators to be sure they are fully aware of all the rules. Membership fees, release forms, and noncompete policies are also common expectations for community radio stations. This, of course, is up to your station management.

After our DJs have taken their first flight, we commemorate the special occasion with a pair of golden WXOX pilot wings.

Addressing Staff Responsibilities

After your ship launches, there is no turning back. You will be out at sea and many obstacles await you. There will be no room for monkey business and all crew members must uphold station expectations. This can be some of the hardest work for the general manager but it is extremely important to make sure everyone is living up to their obligations.

I recommend that stations have all producers and on-air personalities sign a release form and/or volunteer contract outlining the responsibilities and expectations of hosting an FM radio program (as described in Chapters 10 and 11).

Covering scheduled shifts

Upon accepting a radio program on a community radio station, the DJ accepts responsibility for that timeslot each and every week — holidays and special occasions included. Radio never stops!

Passing the broadcast from one DJ to the next is something like carrying the Olympic Torch. If you set it down, it will go out. It is always the DJ's responsibility to cover their shift and to keep the fire burning. However, emergencies occur, so the station should provide resources to assist. Creating a private substitute DJs page on Facebook, for instance, allows DJs to notify one another when they need coverage. An online DJ directory, accessible only to station volunteers, will give them the opportunity to reach out directly by phone or email to other staff members.

Although emergencies are understood, unexcused tardies and absences should not be tolerated. Your DJs must realize that they can never ghost on their show, sign off early, or leave their program unattended. To do so is to set the whole ship off course.

One of the first things new DJs should do is exchange contact details with neighboring program hosts such that they may contact one another if they run into some kind of situation that will make them late or unable to arrive for their scheduled shift. Often those DJs on air before or after them will be able to help out.

When a DJ gets a substitute, they should alert the management. It is always important to know who to contact in times of emergency or broadcast complications. You may want to create an online notification form to keep up with who is on air. If the DJ has underwriting on their show (check out Chapter 8), it is their responsibility to notify the sub and remind them how often the sponsorship message needs to be read.

Holidays

Holiday schedules — including substitutions and prerecorded programs — should be arranged a minimum of one week prior to the event date. Failure to make proper holiday arrangements should not be taken kindly. Goodness knows, station management needs a vacation, too, so they won't appreciate having to be a last-minute stand-in! Luckily, substitutes can be fairly easy to find during holiday seasons as many people break from work and school — and many others seek an excuse to escape the family!

Inclement weather

If bad weather hits and it is dangerous to drive, DJs should be urged to stay safe at home and contact station management as soon as they can. It is often possible to find a sub that lives near the station or who can make it there (see earlier in this chapter for creating a substitute DJ platform). If not, this is a great time to practice remote broadcasting skills (see Chapter 15)!

Sickness and other emergencies

When DJs are sick or otherwise in need of an emergency substitute, they should let the station management know. Management should be compassionate about helping to locate subs in these (hopefully) rare instances and leave the DJ in peace to cope with their personal situation.

CLOCKING VOLUNTEER HOURS

Some stations require DJs to put in a certain number of volunteer hours each month. This can involve alphabetizing records, sorting merch, writing thank you-notes, and more. I've even heard of stations requiring DJs to clean the bathroom on a rotating schedule. What a great idea! At ART FM we ask that DJs volunteer for at least one event per year. This is usually fun stuff such as decorating for holiday parties or working benefit concerts. Nobody seems to complain. In fact, there are lots of folks who go over and beyond, just because they love to be involved with the station.

Each station should develop its own health policy. At WXOX, we ask DJs to stay at home if they are sick and to remain out of the studio for at least 7 days after their symptoms have subsided. COVID-19 has taught us all the importance of being overly cautious when we are not feeling well. At times when serious illnesses are circulating at elevated levels, additional health policies must be signed and upheld by ART FM staff. Anyone who disregards these policies and jeopardizes the health of the staff will face immediate dismissal.

Attending staff meetings

DJs should be required to attend all staff meetings. There will always be updates they need to hear. Full stop. However, you will never find a meeting time that works for all of your programmers. Therefore, it is a good idea to move the meeting around to different days of the week each time. Once a month for one hour should be a good schedule but there will be times when more or less may be appropriate.

It's fine to host meetings in person or online depending on the situation; we mix it up at ART FM. Virtual meetings allow you to demonstrate digital applications but in-person meetings afford you much better team-building capacities. You can generate interest by making the meeting a pot-luck dinner, inviting a guest speaker, or hosting an after-party event at a local establishment. Make your meeting a social affair and you will find everyone looks forward to getting together!

TIP

Because you are busy preparing before the meeting, you will not want to be bombarded with texts from your DJs with excuses as to why they cannot attend. This can be prevented by creating an online form where DJs can submit their regrets. This, along with some kind of roll call or sign-in sheet, helps you keep up with who is putting in the effort to attend meetings. This will be important when it comes to making tough scheduling decisions or dealing with possible HR situations — I find that the individuals not attending meetings will usually be the ones to goof up.

Representing the station on social media

DJs should be encouraged to build social media accounts for their programs on all the major platforms. Through program pages they can promote special episodes, station events, and annual fundraisers. It is a great way for them to connect with their audience on a personal basis.

WARNING

DJs must be careful. Listeners can latch onto voices they hear over the airwaves and sometimes feel they have personal relationships with radio hosts that they do not know. Exercising caution when meeting fans both online and in person is always recommended.

WARNING

Station management should to stay in the loop about what is getting posted on program pages. Although run by independent DJs, these pages reflect on your station. The Internet offers a lot more freedom than the FM frequency; however, any images, remarks, or reposts of a questionable nature should be addressed by station management immediately. Any misuse of social media — discrimination, harassment, slander, or similar — should be grounds for dismissal.

After a DJ is no longer with your station, be sure they remove all references to your frequency or nonprofit organization on their social media sites. The station may want them to sign an exit agreement on the way out asking them to update all their online profiles upon departure.

Cultivating a respectful workplace

Members of community radio stations must uphold the ideals they all wish to amplify within society. They must always respect other people and treat them accordingly. All station staff and guests have the right to work in a safe and respectful environment — in person, online, and over the airwaves.

A community radio station must be committed to eliminating any incarnation of harassment that raises its ugly head. Anyone who observes harassment occurring should have an easy method for making a complaint, and all such complaints must be treated seriously in a prompt and thorough investigation by the general manager or board of directors. The station management must take all necessary steps to bring an end to the harassment including suspension or permanent dismissal from the station staff, and banning someone from the station's premises and activities. Individuals who violate anti-harassment law may also face civil damages or criminal penalties.

WARNING

Harassment is defined by the U.S. Equal Employment Opportunity Commission (EEOC) as unwelcome conduct that is based on race, color, religion, sex, sexual orientation, age, national origin, or genetic information. Harassment can include, for example, racial slurs, offensive or derogatory remarks about a person's race or color, or the display of racially offensive symbols. It was made illegal in the workplace by the Civil Rights Act of 1964.

To clarify any confusion about sexual harassment, the Society for HR Management (`shrm.org`) provides specifically prohibited examples to share with organizational with your staff in the form of an anti-harassment policy: `www.shrm.org/resourcesandtools/tools-and-samples/policies/pages/sexual-harassment-policy.aspx`.

Harassment can take many different forms, from outright attacks to jokes and "playful" touches or comments from coworkers. Harassment may revolve around another person's physical appearance, income, or marital status. None of this is funny on air or behind the scenes. The victim is not always the person being harassed; anyone that witnesses the offensive conduct can be hurt.

Anyone — volunteer or staff member — who is accused of misconduct should be permitted to defend themselves and make an appeal to the full board of directors. If an investigation finds that the report was false, and made with malicious intent, those responsible for the false accusation should face the same disciplinary action. The station should keep complaints and investigations confidential, in order to prevent further embarrassment, discrimination, or retaliation. The board of directors should document all incidents in writing.

Suspension of shows

In order to uphold broadcasting laws and protect the safety of your staff, the station management must reserve the right to suspend or reschedule programs at their sole discretion. Likewise, they must be at liberty to rearrange the program schedule in a way that best serves the community and upholds organizational goals.

For courtesy's sake, DJs should be given two weeks' notice regarding any programming changes. Likewise, the station should be provided with two weeks' notice in the event a DJ decides to discontinue their show. If a program suspension is the direct result of a policy violation, the DJ should not be allowed back on air after they have been dismissed. You must make it clear that shows and program times must be nontransferable; a DJ may not hand their programs over to friends and relations. Everyone must go through the screening and training process that keeps your studio, your staff, and your frequency safe.

5

Riding the Waves: Ups and Downs in Community Broadcasting

IN THIS PART . . .

Embrace the highs, understand the lows, and buckle up for a wild ride in broadcasting!

Seize opportunities beyond the studio by building local relationships and embracing your role in your community.

Connect with listeners internationally, broadcast remotely, and syndicate your programs worldwide.

- » Dealing with people problems and human error
- » Bouncing back from technical difficulties
- » Expecting the unexpected

Chapter 13
Understanding Murphy's Law of Radio

Everything that can go wrong at a LPFM radio station will go wrong at yours! You will have to be a radio ninja and be prepared for every situation. Just when you think you have all the bases covered, the thing you never imagined will suddenly occur.

But don't worry, this chapter is here to help.

Staying calm and collected in the most exasperating situations is much easier said than done. ("I can't believe this is happening again — right now!"). But as the station's leader you must breathe in, close your eyes, focus your mission, remember why you are doing all this, and breathe out. Breathe in. Breathe out. Focus. Repeat. And then open your eyes and solve the problem.

There is no downtime in radio. You are always on stage. If something has to get fixed, everyone is waiting for you to fix it. Impatient ears are listening. Friends are texting and DJs want to know what is going on. This is the one and only time when you will have to tune them out, forget about what the listeners are thinking, and tune into the problem at hand.

TIP

An adventurous audience can enjoy the troubleshooting experience. When things are sounding awkward on air it can be a good opportunity to remind the listener just how special you are, "now that's something you won't hear on commercial radio." Stay positive because oddly enough, we at WXOX always get a boost in listeners after we go off air. Stay silent long enough and we even get donations! It is sometimes good to remind your listeners exactly what it is they are missing.

TIP

In the online resources for this book, you can find a technical difficulties log for documenting when things go wrong and how you fixed them. Make this accessible to your team online so that staff can see how things were fixed when the same problems arise again (and they will). Although you are not required to document most of these technical issues, the Federal Communications Commission (FCC) does require you to keep track of any time the transmitter is turned off and when it was turned back on. This spreadsheet is a good place to keep that information.

This chapter is about learning from your mistakes (or learning from my mistakes), and recognizing that every wrong turn is a new discovery. I can tell you everything I know, but ultimately you will have to learn much of it the hard way yourself. The most important thing is not to sweat it. You'll get everything fixed — eventually — and your listeners will forget that the station ever went off air in the first place.

Being Prepared When the Broadcast Goes Silent

Ironically, every time I've sat down to write this chapter, WXOX has gone off air in some way or another. The first time I thought it was something I had done. I was logged into the station computer uploading a program special when suddenly the station went off air. It was almost 2 a.m.; I was feeling sleepy, so I could have messed up. Yet, I could not for the life of me figure out what I could have done wrong. So, I got in my car and headed down to the station. It was a rainy night, with flooded roads, and I was halfway there when fresh out of nowhere the signal returned. Loud and clear. The station was back on-air. I turned around and went to bed; all's well that ends well, I thought and had a good night's sleep. Deep from my dream about sunbathing on an Italian clifftop, I was startled into reality by the sound of rattling windows at home. It was fiercely windy but the sun was shining and the birds were singing. I stretched, pulled myself out of bed, eased into my bunny slippers, and switched on the radio. Silence. I rubbed my eyes and stumbled over to the computer. More silence. I remembered that it was April Fool's Day. "Not funny," I said out loud, and then the troubleshooting began.

The following sections look at some of the most common problems you'll encounter and how to address them.

User error

Always suspect user error first. When you think the station is off air, the first thing to check is your listening device. Make sure you aren't doing something wrong. Then, try using another device. If the other device is working, great! Problem solved (at least from the station's standpoint).

If the audio is still not working you now need to check another type of device to see if the web stream is down, the FM, or both. If you discover that both are down, it is most likely a studio problem; you can skip to the section "Failing Equipment," later in this chapter.

Utility failure

REMEMBER

When your signal is completely off air, start with the question so basic that it seems stupid to ask: "Is it plugged in?" After you've verified that devices are in fact plugged in, ask: "Is the electricity on?" If not, and you don't have a generator, there is not much you can do but wait — and think about how much you wish you'd invested in solar!

If the electricity is working, next you should ask: "Is the Internet working?" Depending on your setup, you may be able to keep audio going to your FM tower without Internet service, but in many cases where the tower is fed audio remotely, you cannot.

Your web stream cannot work without access to the Internet. If your Internet connection is not working, the first thing to do is to reset your router and modem (including disconnecting all the cords and reinserting them securely). Plug your modem in first and allow it to fully boot-up — with all lights illuminated — before plugging in your router. It often helps to restart your computer after this reboot process is complete, to get a fresh connection. If you are still not able to access the Internet, there is a problem with your service or your equipment, and you should check with your provider to see if there is an area outage or if the line needs attention.

TECHNICAL STUFF

Wi-Fi hotspots generated by mobile phones can be used temporarily during times when there is no standard Internet available. Most wireless phone services offer this hotspot option (some for free, others as a paid upgrade). Hotspots are very handy but often come with significant charges if you go over the time or data limits so be sure to keep up.

Radio silence

When electricity and utility services are working but you are still experiencing radio silence, immediately check all devices and connections between your soundboard and your FM transmitter.

If everything is connected correctly, you need to do some deeper digging. Could a cord be broken? Or is it a piece of equipment that's malfunctioning? FM transmitters can get fried by lightning strikes and other bad luck.

Money for replacing these essential broadcast items should be allocated to a rainy-day fund as soon as your organization is able to set one up (for more on planning for station expenses, check out Chapter 9).

Streaming silence

When the *stream* (the online broadcast) goes silent, troubleshooting becomes a little more complicated. The first thing to do is to check your broadcast software. Does it say "on air"? Are the levels looking good, and are the counters moving? If yes, and everything seems to be working correctly, the problem may be with the code of the website that makes the player button work.

In most cases when your stream is not working, the encoding software will let the broadcaster know with an error message of some kind; this varies between different software but it should be bold and noticeable. With any luck this error message, like the one provided by the audio-encoding software Audio Hijack (`rogueamoeba.com/audiohijack`) in Figure 13-1, will tell you exactly what is wrong. The following sections tackle some common problems.

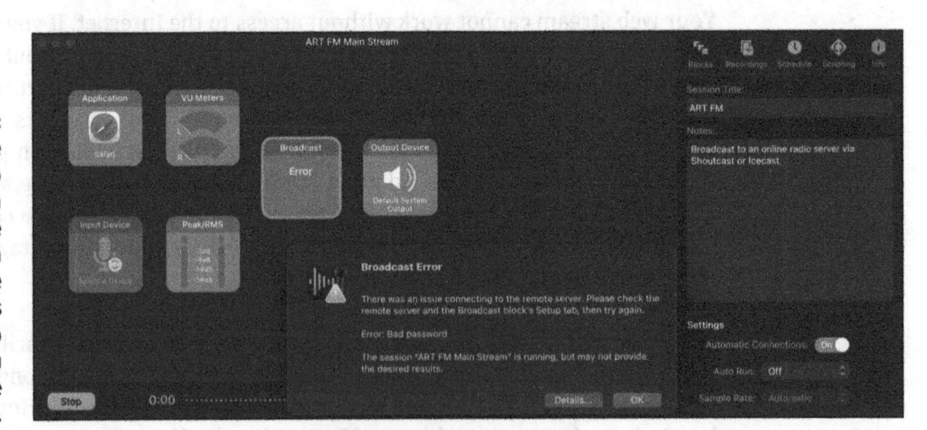

FIGURE 13-1: An error message shown in Audio Hijack. Luckily, in this case, the program identifies the problem and it is an easy fix (so long as you remember the password).

WRONG ACCESS CODES

It is possible that you messed up the *stream access codes* (the codes provided by your stream host for connecting with your URL) if you have changed or updated your broadcast software recently. You may have typed something wrong into the encoding software setup. Look over your broadcast block with a fine-toothed comb and compare it to the access codes provided by your stream provider. Even one small typo will make it impossible to connect. See Chapter 5 for more on making stream connections work.

ANOTHER USER ON YOUR STREAM

Do you allow anyone else to connect to your stream? For instance, a remote broadcaster or someone in an alternative studio? If so, and you cannot connect, make sure that they are not connected to the stream. Because only one computer can be connected to the server at a time, if they are broadcasting (even inadvertently) you will not be able to connect.

STREAM SERVER HAS CRASHED

It is possible that the stream server has crashed but you will not know until you or another member of your admin team has logged into the control panel. If the server has crashed, you'll need to push the Start button in the control panel to get it restarted.

Most of the time this happens when the streaming computer tries connecting to the host with too high a bit rate (for more on the technical aspects of streaming, head to Chapter 5). Before reconnecting, check that the bit rate on your streaming software is set to the correct speed.

If your server has crashed and you are unable to restart it, you will need to contact your stream host provider. There may be an outage on their end. If this happens with any frequency, you should seek a new provider because it is critical that your stream server is stable.

Dealing with Static and Other Interference

Sometimes everything seems to be working fine but the audio signal just sounds bad. It may be glitchy, heavy on static, garbled, or just really quiet. This is caused by audio interference, which can affect both FM and Internet broadcasts. There are infinite reasons why sound may become distorted, and as you dig into this unruly medium you will begin to understand why audio engineers demand big bucks!

As always, with engineering, start with the easiest solution and work your way deeper. Squelches, squeaks, static, and interference can come from any number of sources. Fingers crossed that it is just a cable and not your soundboard or something major (and expensive).

The first thing to do is to isolate the problem. Is it on the FM only? Or is it on the webcast? Or is it system wide? Check this by tuning into the stream and the FM signal independently to see if you get the distorted sound.

If it is system wide, start by looking for problems in studio. Did you test the connections? Always try different channels, cables, and connectors before you blame the equipment. How are the audio levels? Too high "in the red" and the audio will distort; too low and no one will be able to hear it.

FM interference

Interference with your FM signal often comes from other radio broadcasts. This is why the FCC issues licenses so that, hypothetically speaking, two FM broadcasters do not interfere with one another. However, radio transmission depends on atmospheric conditions, which are not always predictable. When an FM broadcast sounds bad in areas where it normally sounds good, you can blame it on the "tropo" (short for tropospheric ducting).

It's time to don a white lab coat for this explanation! Scientifically speaking, *tropospheric ducting* occurs when a radio signal encounters an atmospheric rise in temperature at higher altitudes rather than the expected cooling. The warmer atmosphere causes the radio signal to bend and reflect. In turn, this can cause signals to carry great distances. Along the way, this causes interference to stations broadcasting within their designated airspace.

You cannot see tropospheric ducting when it occurs but you sure can hear it. In Louisville, we often get the shock-jock programming from Hank FM in Indianapolis interrupting the WXOX signal; we can hear their station identifiers as clear as day. And I promise it was an unsettling holiday surprise when the Hank DJs started singing "Candy the Red-Nosed Stripper" over an arts program on our community radio frequency, but there was nothing we could do about it. Especially since LPFM stations are presently not protected from interference by the FCC.

To confirm that the interference you are hearing is tropospheric ducting, you can consult the VHF propagation map online at `vhf.dxview.org`. It will be obvious if there is activity in your area, as shown in Figure 13-2.

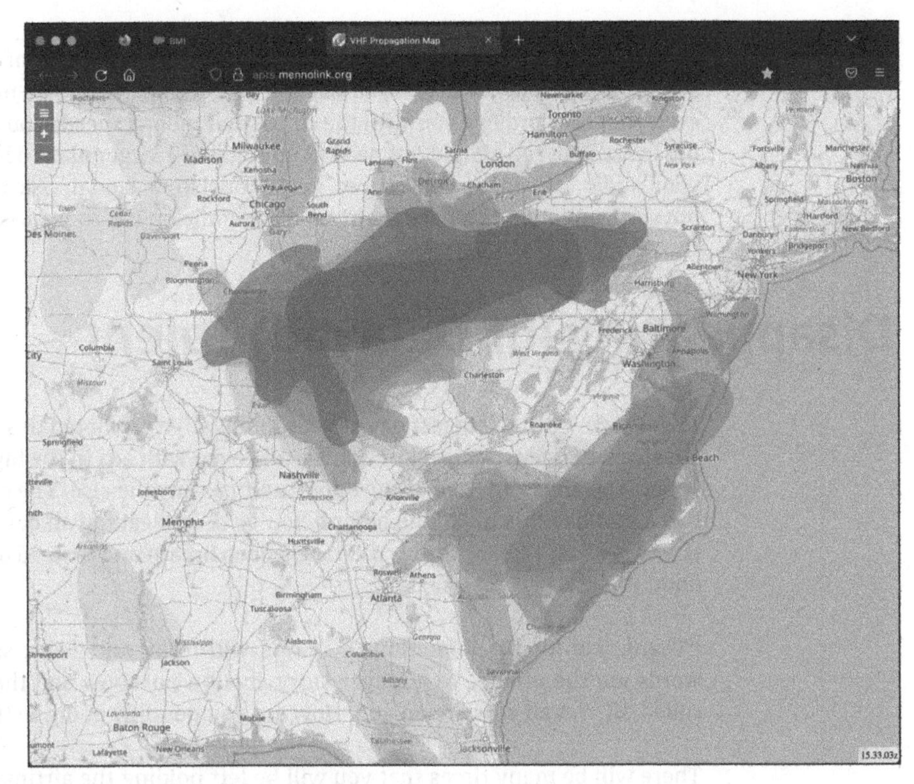

FIGURE 13-2: The VHF propagation map from `vhf.dxview.org`. This site displays tropospheric conditions in real time with updates every minute.

Stream glitching

There will be times when your station's web stream becomes *glitchy*. There is really no other way to describe it. The sound will come in and out and add some unique digital touches.

If you have problems with a glitchy stream, make sure that there is no other traffic on your Internet system. Uploading videos can be particularly problematic, as can folks on other devices playing video games through the same connection that you're using for the webcast. Make sure that all your software — including your broadcast software — is updated and that you don't have any extraneous programs running in the background.

TIP

If the stream continues to glitch, you will need to check all of your connections. Make sure you are plugged in with an ethernet cable (see Chapter 5) and not relying on Wi-Fi. Try unplugging and reinserting the connection again. Note that sometimes Wi-Fi gets turned on and overrides the ethernet connection on the computer. So make sure that your ethernet is plugged in and your Wi-Fi is turned off. If the glitching continues, reboot your system, beginning with the modem and router that provides your Internet service.

In most cases this will take care of it. If not, you may have to contact your Internet service provider. They may try to sell you a more expensive data plan. Insist, however, that they provide you with the consistent upload speeds you are already paying to receive. You should not need to upgrade to a larger plan to handle uploading your audio stream. If you cannot get a consistent signal, ask them to test the physical equipment or have them send a technician out to inspect the line.

Disappearing (or Delayed) Talent

It will seem like the biggest day of your life! You've promoted this event online, on air, through flyers on telephone poles, and even with ads in the high school newspaper. You are as amazed at the opportunity as anyone else. The station has never had a star like this in-studio before! You've been cleaning all day, and everything is just perfect. You are ready. You take a deep breath and instead of a knock on the door, it is the beep of a text.

"Thank you for the invitation . . ." And with that your heart sinks. These five words are the prelude to a longer story about a no-show and the whole thing is called off. Yet all ears are on you. How are you possibly going to fill the air?

There will be many times that you will be left holding the airtime that somebody else drops. Sometimes it will be because someone else is remiss; at other times, it is because of forces beyond anyone's control. Either way, the result is the same and it is time for you to head quickly to Chapter 16 for some timely inspiration.

Failing Equipment

Things break. They just do. Some more frequently than others. It comes with the territory in community radio so you just have to expect it.

Training your recruits to treat the equipment well is important, but invariably accidents happen — and with a lot of use, from a whole lot of people, things also just wear out.

Turntable needles, for example, are super fragile. You will need to have plenty on hand. You will know it's time to change the needle (or *stylus*) when records are sounding fuzzy or not playing smoothly, even after the needle and the record have been cleaned. Cables that are in heavy use have a tendency to go bad over time, so have at least two or three extras in-house at all times. Channels will die out in headphones, so you will need extra sets of these as well. If the problem is deeper, hopefully you have a local repair shop in town that can assess your gear and help you to isolate vexing problems without having to purchase all new electronics.

You should have backup for any equipment — computers included — that you just can't broadcast without. If it's not in-house, make sure you have the means to get essential equipment replaced and installed quickly. Plan ahead. Every minute off air will feel like a lifetime.

Remind your volunteers that they should never, not under any circumstances, adjust the software settings or the equipment setup on their own. This will greatly perturb the engineering team at your station, who have likely spent many hours getting everything tuned in just right. As soon as one person adjusts something in the studio, others seem to follow suit. It doesn't take long for the whole thing to get way out of whack.

Spilling onto the Soundboard

Beverages will spill on your soundboard if you allow them in studio. It is not if, but when. And this is bad news — very bad news — for the electronic equipment you use when broadcasting.

As a result, beverages (and food) should not be allowed in your on-air studio. Not ever. That is my opinion and ART FM DJs know that they will be dismissed on spot for violating this rule! We have a countertop just outside the studio where food and drinks can be shelved. When someone needs a bite or a sip, they can easily step just outside the studio and access something tasty or refreshing.

Breaking the Rules

You can have meetings, send documents, post signs, and send messages but no matter what rules you put in place, they are bound to get broken. I have no idea how to put this differently or to change this reality. It's just how things are. Therefore, prepare yourself mentally. As a leader of a LPFM station you need to be able to handle rule-breaking situations fairly and evenly for all members of your staff.

The general manager of the station must agree to enforce existing rules and should not make any new rules that they are not willing to uphold. Sometimes enforcement means a slap on the wrist, other times it's outright expulsion. In most cases, it is the job of the general manager and program director to uphold station policy among the staff. Serious human resources cases should be taken to the board of directors who will vote on disciplinary actions. (Check out Chapter 12 for more on these roles and for handling discipline within the radio station, and the next section for more on looking after yourself.)

TIP

Should there be slips and FCC rules (see Chapter 10) are broken by your staff, my advice is as follows:

1. **Contact the DJ immediately.**
2. **If you cannot contact them or if they are anything but apologetic, take their program off air immediately.** You should be able to do this remotely in many cases. Do not let the DJ back on air until you have spoken with them and you feel confident that they understand the gravity of the situation and will not make the same mistake again.
3. **If the DJ is apologetic and it was clearly a mistake, you can issue a warning and cautiously allow them to carry on.** At ART FM, first they get a warning; then a two-week suspension if it happens again; then full suspension of their show and removal from the schedule on the third strike. This is serious stuff! Presenters must understand that FCC violations can result in fines up to $10,000 per instance. One song can bring your nonprofit organization total financial ruin.

REMEMBER

This prospect should not discourage you from pursuing your radio dreams. FCC law is extremely important and your station's responsibility is to uphold it at all times. When a mistake is made — and it will be — you don't need to freak out. You instead need to respond appropriately and then keep calm and carry on.

If a DJ makes an FCC or station policy mistake, they should not draw additional attention to the matter but rather move forward with the broadcast. They also should not mention the mistake to anyone but the station management and they should not share any unedited recordings of the episode. Some stations keep a *black book* of any station FCC violations. This details how they happened, by whom, at what time, and how the issue was addressed by the management in order to prevent it from happening again. This book should be kept private, accessed only by general managers and the board of directors.

WARNING

The FCC takes complaints filed against FM radio stations seriously. All complaints with evidence of wrongdoing will be investigated by the agency. For more on FCC law, head to Chapter 10.

Ken Freedman of WFMU recounts that this miserable experience includes people in suits armed with clipboards sitting in the studio taking notes of everything the DJs said and played on air. Ken now calls his radio program *Seven Second Delay* in homage to the devices that can be installed in the transmission chain that delay the signal by seven seconds so that time can be cut out before hitting the FM airwaves, if necessary. This device is called a *broadcast delay* or *dump button* and costs around $1,500, but it is a piece of equipment that you may want to consider. Remember, however, this will put your FM broadcast on a short delay and may not be helpful in instances where a cuss word is repeated multiple times.

Staying Tough and Carrying On

There will always be problems — more than you can ever possibly imagine — and you can never prepare for everything. As you face these challenges, I hope you will be able to laugh at how absurd things can become. The most important thing in radio is being able to think quickly and creatively. You will be surprised at what you can come up with, when necessity makes you. Remember, there are no mistakes only new opportunities!

Shepard Fairey, contemporary artist and founder of OBEY Clothing, was discussing street art in an interview on ART FM recently when he said, "If your feelings get hurt easily, you are not going to last."

His words resonated for us at ART FM when everything started going wrong during a trifecta of OBEY broadcasting events that included an interview, an artist talk, and a live DJ set. Looking at the happy photographs in Figure 13-3 you would never know that the stream died, the audio failed, and someone messed with our sound settings! Despite everything going south, it still ended up being one of our most epic broadcasting accomplishments to date.

If there is a moral to this story, it is to make room for important broadcasts and cut yourself some slack when they happen. Things may not go perfectly. In fact, everything that can go wrong will go wrong. But if you just try, you'll discover some really great things will happen as well.

Had our feelings been easily hurt, our emotions easily embarrassed, or our resilience been too low when working on these OBEY broadcasts, we never could have pushed through all the trials and emerged with such a great feeling of accomplishment.

FIGURE 13-3: Everything that could go wrong, did go wrong for WXOX during this series of OBEY broadcasting events, which included an interview, an artist talk, and a live DJ set from Shepard Fairey. At the same time, everything kept going right. Each opportunity led to another and listeners from all over the world were letting us know that they were tuned in!

© Sharon M. Scott (Book Author).

In this business of radio, it is important to stay tough and remember what a good thing it is that you are doing for the world. Folks will blame you for things that they know are not your fault. They will expect the world of you and then show very little appreciation when you deliver. It can be very draining. Remember that although you will have many allies, there will also be those who are looking to trip you up out of jealousy or spite. They will wish they were in control of the radio station and — with no basis in reality — will imagine that they could run it better. Women will encounter sexism. People of color will face racism. Older people will face ageism. The list goes on. No matter what makes you different from everyone else in the world, someone will try to turn it against you.

In your wisdom, however, you know this difference is the very thing that gives you the strength, the creativity, and the tenacity to carry on. Focus on the real challenges of building a valuable media resource in your community and you will not have time to get caught up in the nonsense.

WARNING

At times when you need to discipline a DJ or suspend a program, be very careful. This can be a very painful experience for everyone involved. I have learned the hard way how important it is to do your diligence on the front end. Radio programs become a part of a person's identity. When they feel that this part of their identity is getting taken away, there is the potential for them to act irrationally. Social media backlash from former staff members is a common occurrence in any business. When they happen, remember you are not alone. Act honestly and fairly on behalf of the station and you can be confident that these empty statements will evaporate quickly enough. Focus on cultivating your station and soon you will be so busy that you won't even recall how painful it was at the time. Check out Chapter 12 for more advice on handling delicate matters within the station.

Non-disclosure and exit agreements signed when onboarding can proactively prevent retaliatory behavior. When a DJ leaves, they can be reminded of the agreement they signed. If online retaliation occurs, the station and its managers should not engage but seek the advice of the board. Remember that anything you say or write can and will be used against you. Consult an attorney if any social media posting or personal messaging becomes threatening to you or your nonprofit operations.

REMEMBER

Be careful to protect your mental health at all times. If you are the station's general manager, the buck stops with you which makes a lot of weight to carry on your shoulders all day, every day. Make sure you know your limits, and delegate suitable responsibilities to other members of staff. Put in place a response team that can handle emergency situations when you need some time away. You simply will not be able to do it all yourself. Strength lies in numbers.

IN THIS CHAPTER

» Developing relationships through interviews and events

» Becoming a resource for your community

» Preparing your local emergency response

Chapter 14

Cultivating Community Connections

Friendships will be made, collaborations will take flight, and romance will set it all aflame! By definition Low Power FM stations are community meeting places. They are sources of energy and conversation, and many amazing things will happen within your station walls — but not if it is an echo chamber. Nothing good comes from going it alone. Making allegiances, building partnerships, and providing coverage of local activities are key to developing interesting programming for your station and engaging listeners with your broadcast.

LPFM stations use their creativity as you can see in Figure 14-1 to develop new ways for connecting with their listeners, which includes everything from record fairs to costume parties. This chapter provides ideas for connecting with your community, recruiting diverse voices, documenting important local events, and being on call for your community if disaster strikes.

FIGURE 14-1: Beyond their broadcast, LPFM stations create all kinds of happenings to connect with their communities.

Bringing the Community to the Station

Your LPFM radio station should reflect the culture, interests, and personalities of your community. The best way to do this is to invite the community inside. Open your doors and people will show up to see what is happening!

Hosting open studio events

Open studio events can be one of the easiest ways to get to know your neighbors, and folks enjoy a peek behind the curtain. Invite the DJs to participate, too, and your open studios become fun, low-key gatherings. You simply pick up a few refreshments, get the word out, and that's it! You've already got the entertainment!

When ART FM was situated in the Nulu downtown artist district of Louisville we hosted monthly open studio events in coordination with the neighborhood's First Friday open house activities. This was a great way for us to make an entry into the local arts scene. Since then, we have enjoyed open studio events to connect with area creatives and recruit new members of the team.

Interviewing great guests

One of the best ways to get community members involved with your radio station is to invite them to share their interests on air. It is really fun for folks to see the studio and it is easy to sell an FM radio opportunity.

WARNING

Be sure to review the Federal Communications Commission (FCC) rules in Chapter 10 before embarking on hosting any on-air guests, as there are specific rules that may apply to each type of interview.

TIP

From musicians to politicians, you want to keep a revolving door of local characters appearing on your broadcast. Check out the following list for great interviewee ideas:

- **Nonprofit organizations.** Local nonprofits are an excellent source of good conversation about relevant topics. Most of them will have a director or PR person who is ready to talk about their current projects at the drop of a hat. Arts organizations with upcoming exhibitions or performances are always enthusiastic about getting the word out about their events. They can likely bring some additional interviewees such as featured visual artists, actors, or musicians. Performers from local theater companies are always itching for a spotlight and may even want to produce some radio-specific plays.

  Just remember that although you can and should promote the nonprofit activities in your community, you cannot fundraise on air for any other organizations except the LPFM license holder. This includes encouraging ticket sales and the like. (See Chapter 10 for more on these FCC rules.)
- **Local businesses.** Smart businesses are always interested in the publicity that comes with being on the radio. You are welcome to interview business owners but just keep the FCC rules in mind. In noncommercial radio it is important that the host remain neutral during the interview and not make any endorsements or calls to action. This is especially true in cases where the station is receiving any sort of financial gift or donation from the business entity, which would be a serious FCC violation.
- **Celebrities.** Who are the celebrities in your community? They may be artists, musicians, doctors, first responders, teachers, charity workers, or anyone else worthy of your station's attention. No matter how small a town you live in, there are people around you making great things happen. From the arachnid expert at the local nature center to the kalimba player at City Hall, there are so many interesting people that the world should notice and celebrate. Seek out these individuals and ask them to come share their talents on air.

TIP

Getting interviews with bigger celebrities who are traveling through town is also very exciting. To score these interviews you will need to reach out to the event's promoter. First, seek a press contact on the event's website or online listings. If this doesn't work, contact the venue, making sure to note that you are an FM radio station seeking an interview, and they should be able to put you in touch.

- **Festivals and larger events.** Most events will likely have a press page on their website that provides you with the means for applying for press access. If you request such permission, remember to use it in a professional manner by agreeing to provide a report or review from the performance either on air or in a station publication. Send a copy of any recordings or materials to the event organizers afterward. They will appreciate a copy for their files and remember it the next time you make a request.

  When you request a press pass you may be able to request give-away tickets for your listeners while you are at it, but remember, you must run these contests fairly (see Chapter 10). You must avoid personal gain; payola is strictly prohibited by the FCC rules.

- **Politicians.** Political characters can be among the more difficult interviews to wrangle. Often, they would rather say "No" to an interview than risk saying the wrong thing on air. You will need to request an interview through their website, which I recommend following up with an introductory phone call to the office.

REMEMBER

Specific FCC laws exist pertaining to equal airtime for political candidates who are actively running for office. Likewise, noncommercial radio hosts may not endorse any political candidates or weigh in on any legislation that is coming up for vote. You must absolutely review these rules in Chapter 10 before embarking on any political discussions on air.

Getting past the gates

TIP

Interviews with politicians, athletes, and A-List entertainers can be a challenge to secure. Here are a few tips for getting past the gatekeepers:

- Build a reputation of professionalism and objectivity. You can do this one interview at a time. Be aware of your digital footprint and create an online presence with documentation of your recent work.
- Submit a one-page reporter profile with your request that includes education, credentials, awards, interests, and possible testimonials. Include contact information and links to social media.

- Use your networks. If you know a person or an organization with a firsthand connection to the interview subject encourage them to reach out on your behalf.
- Be polite; be persistent; be creative.

Securing a release form

After an individual agrees to an interview, your station should ask them to sign a release form before going on air. Your release form needs to provide your station with the necessary rights to broadcast, rebroadcast, record, and distribute the individual's voice. (You can learn more about release forms in Chapter 11.)

Sometimes a release form is not necessary, such as when it is obvious that the person knows their voice is going over the air (for example, when participating in a call-in talk show). It can also work to record them saying that they acknowledge the recording and give permission for it to be shared. However, you should always consult your own attorney in regard to obtaining legal permissions.

TIP

At ART FM we provide our guests with a single sheet of information that helps them prepare for the interview. It includes directions to the station, parking information, and a request to arrive 15 minutes early. Health and safety information is included as is a reminder to uphold FCC rules. Website and social media information is also provided and the guest is kindly asked to help spread the word. You'd be amazed at how often folks won't bother to post about their radio appearance unless you specifically ask them. You can find an example guest info sheet in this book's online resources.

Conducting the interview

When it comes to recording the interview, make your guests feel at home. Whether you are interviewing them on a busy street or in the comfort of the studio, you want them to feel respected and comfortable while speaking with you.

TIP

Here are a few tips to help set your interviewees at ease:

- Pronounce the name of your guest correctly. Practice it with them before going on air.
- Know their official title and the name of the organization they represent.
- Have the guest's website, event details, and other relevant contact information ready to share.

» Announce your guest periodically in the conversation and after each break, for listeners just tuning in.

» Keep quiet while your guest is talking. Turn off your mic if necessary. Do not add audio comments to the conversation: No one likes hearing "umm-hum," "yes," "ah-ha," or similar sounds. Instead, provide visual responses and encouragement.

» Do your research. Have unique, insightful questions prepared.

» Have some relevant music ready to play. Perhaps from their group, perhaps something they recommend. Just make sure it is FCC friendly!

In most interview situations there's some lighthearted banter and some serious discussion. It is nice to strike a balance between the two. Make sure the music you choose and your general tone is reflective of the mood of the conversation. When getting into the discussion, always remember to include your listener as a participant in the conversation.

Reporting on community trauma

It's important for your station to be there for the community in times of tragedy and upheaval. Dealing with these subjects requires a lot of sensitivity and it is important to always be respectful of your guests and understanding of where they are coming from. If someone has recently been through an intense situation, perhaps there is a trauma counselor you can invite on the show as well. This may be helpful in case things get beyond your realm of expertise — though sometimes people just need to share their stories.

TIP

Make sure to provide plenty of musical breaks — while being sensitive about musical choices — so you have an opportunity to speak with your guest off air, to make sure they are feeling okay with the direction of the conversation. Have tissues available in studio (for you and for your guests), and have a way for them to signal when they need a break. Also be able to provide your guests and listeners with websites, phone numbers, and community resources that can help them cope with the struggles at hand.

Bringing power to the people!

Traditionally, the power of radio has been inaccessible to many communities. With LPFM stations, we can turn this tide and present opportunities to those who have not had them before. When considering interview guests, look to amplify the voices of women, minorities, immigrants, and speakers of different languages. These people are stakeholders in our community, and it can be enlightening to find out what their reactions are to recent legislation, community activities, and neighborhood projects.

People of all ages, from young children to those in their golden years, should be welcome on your broadcast. Let them use the airwaves to build bridges (see the later section, "Telling veterans' stories," to see how WXOX teenagers are connecting with local retirees). Make sure people of all cultural backgrounds have an opportunity to share their perspectives, which are sometimes overlooked by mainstream media.

TIP

If you host an interview program and can't think of another topic to cover, why not explore another culture? It is always interesting to learn about the language, festivities, music, and delicacies belonging to other parts of the world.

The Freedom Hour on ART FM is a weekly program hosted by Nilou Sabzevari, Shadi Kavoosi, and Shirin Bolhari. Although the primary focus is on the Iranian Revolution, their show includes some lighthearted segments including lessons on the Farsi language and popular expressions. For example, I recently learned that "You can't lift two watermelons with one arm." Who can't appreciate this kind of universal wisdom?

Working with youth groups

Having school and community youth groups visit the station is always so inspiring. In addition to a tour, give them a moment on air. Who knows when else they will have the opportunity, and I promise they will never forget it!

Of course, depending on the size of this group this can be unruly, and you have to keep your listeners in mind. Instead of allowing the youngsters to jump on the mic, giggle, say "Hi Mom" and the like, you may want to have prompt cards handy. These should include sentences they can complete, such as "Someone I consider a Superhero is . . ." or "Something I'd like to do for the world is . . . ." The prompts get them accustomed to speaking on air without having to think of topics to cover. That can always come later, after they settle in.

Telling veterans' stories

It is an exceptional honor to have U.S. combat veterans in the studio to share their war experiences. The Veterans History Project (VHP) at the Library of Congress collects, preserves, and makes accessible the firsthand recollections of U.S. military veterans (`loc.gov/programs/veterans-history-project`). If your station records interviews with local veterans, you can also submit your recordings and develop a VHP collection in your station's name. The primary requirements are that the subject be a veteran of combat and that the recordings be a minimum of 30 minutes. Signatures from veterans or their families must be provided on certain forms. Photographs and other items such as letters or articles may be included within a veteran's file.

Your radio station can be a great resource for recording and preserving their memories of defending American freedom. The Library of Congress offers field kits complete with tips and helpful information to help preserve veterans' histories. Staff librarians can even host a virtual workshop for your station and other interested parties. If you have questions or would like to schedule an in-person visit to the collection, email `vohp@loc.gov` or call 888-371-5848. The administrators are very enthusiastic about this project and ready to help your station get a collection started!

At WXOX we are lucky to have World War II Navy veteran George Fox hosting the *Opera Hour* every Monday. During one special episode, he shared his harrowing war experiences with our listeners — including his assignment in Hiroshima Bay in 1945. His story was so compelling that teen director Max Selby contacted the VHP and started a collection on the station's behalf. Veterans are now invited to connect with ART FM through our website (`artxfm.com/vets`) and schedule an opportunity to share their stories on air to the teenagers in our *Next Wave* program. The students record the interview as it happens and subsequently prepare it for submission to the VHP collection.

Organizing DJ training programs

As a community radio station, you need to teach your community how to create radio broadcasts. Each station develops its own method for training local volunteers on the technology and FCC rules.

It's a good idea to host an introductory community meeting in which you provide information — but no promises — about your DJ training program and how individuals are selected for shows. This meeting may be in person or online, possibly both. It is an excellent way to give interested parties an introduction to your station and an understanding of program responsibilities. It is also a good opportunity for you to meet the applicants in person, to get a feeling for the type of energy they will bring to your organization. See Chapter 12 for more information about training and onboarding new recruits.

Offering internships

Creating an internship program with educational institutions in your community can be as easy as sending an email. Colleges and high schools, particularly those with media departments, are always seeking real-world opportunities for their students. The energy and enthusiasm of young volunteers can be a great asset to your station. Unlike a big corporate or public radio station, you will likely be able to offer the interns management experience and on-air opportunities that will help them professionally — whether or not they pursue a career in radio.

When you accept interns, make sure to engage them in activities that will benefit their studies. Don't give them all the grunt work. Helping with remotes, creating zines, and building social media are all great ways to get interns plugged in. If you have the airtime, I highly recommend getting them trained to host a show. Some of our best programs have come from our interns and this experience can offer an excellent jump-start for their future career.

Riding the Next Wave: Teen Radio Academy

Teenagers love radio if you just give them a chance! To introduce them to the art of broadcasting, you may want to consider hosting a summer radio academy.

There are lots of activities to keep younger kids occupied during the summer break but not so many for teens. To fill the gap, ART FM established the week-long, half-day Next Wave: Teen Radio Academy. The Academy gives teenagers the tools and FCC training they need to become professional broadcasters (find out more at `artxfm.com/nextwave`). By the end of the week, each student gets the opportunity to host their own half-hour radio program. Many of them have stayed involved with the station since getting a taste for radio.

It turns out everyone wants to be heard — especially teenagers!

ENGAGING THE YOUTH

Max Selby, The Next Wave Director at WXOX, shared the following insight on encouraging younger members of your community to join in at your LPFM station:

"To stay relevant with younger audiences, radio stations must bring in diverse DJs! Coming from a teenager, trust me when I say no one my age wants to listen to a twenty-four-seven rotation of middle-aged men playing the blues. It is crucial to have a diverse roster in not only background but also in age. Young people will listen if they find that they can relate with the voice behind the mic, and this is impossible when all of the DJs are similar. At our radio station, WXOX, we have recently brought some younger folks on deck, and it has certainly boosted many of my peers' interest. What works is when someone who knows what the youth of today are listening to. Getting someone on air who is familiar with the local music scene to play upcoming bands or trending TikTok

(continued)

(continued)

songs is vital. Although they have a lot on their plate, high school and college students make great DJs. On many occasions I have been scrolling through social media and have seen young people post about our station playing a song they like, and it is almost always during a 'hip' host's show. So basically, bring in young DJs and you will bring in young listeners.

Social media is going to be your best friend when trying to recruit younger DJs. Simply posting a call for young DJs on Instagram or Tiktok is sure to attract some candidates. Using relevant hashtags and asking your staff to repost will only widen your circles. Another great way to find younger DJs is to reach out to local high schools. Offering internships, summer camps, or after-school opportunities is a win-win. High schoolers want any platform they can get activated, and radio stations should want every teenager they can get involved."

Bringing the Station to the Community

Halloween parades, sporting competitions, and theater festivals are a few of the lively events that ART FM has enjoyed participating in over the years. There are so many interesting ways to connect with your community through events of your own invention or by participating in those organized by others. No matter how you go about it, reaching out to your neighbors and introducing them to your radio broadcast is really important. That personal connection makes all of the difference. It will often be the reason they take the time to tune in, and is the reason they tell a friend about your station.

Taking part in parades and festivals

Neighborhood parades are a lot of fun — both for your staff and your community! They are highly creative, low-stress opportunities for your station to make a display of itself. Participating in parades can be as simple as adding streamers to some bicycles or as complex as building a full-on mechanical float!

The staff of KMRD marches with colorful handmade standards in the Christmas and Fourth of July parades in Madrid, New Mexico; WFAQ turns their vintage BMWs into fully outfitted sound machines for an annual parade of antique cars in Mukwonago, Wisconsin. There are all kinds of ways to make a spectacle of your station.

ART FM built a boombox float for the annual Pegasus Parade at the Kentucky Derby Festival. This event is unique to Louisville, but is an example of local opportunities that will benefit from your station's involvement. And vice versa!

TIP

Hosting information booths at arts events, local festivals, and other gatherings is always a fun and easy way to connect with the community. You can see an example of WAYO Rochester's tabling setup in Figure 14-2. The main purpose of tabling at these events is to let folks know about your station and how they can get involved — as DJs, listeners, or supporters. To participate in these community events, you should have an event box ready to roll that includes the following:

- Tablecloths: Two or more (one branded with your station's logo)
- Signage: Banner and free-standing signage
- Stickers: Branded with your call letters, frequency, and website
- Postcards and brochures: Giving information about your station
- Donation information: Forms and a donation jar
- Email sign-up: A clipboard list for interested parties

For more on station merchandise, see Chapter 8.

You will also want a sound system, of course! This may be a portable radio or a full-on sound system, depending on the nature of the event and your role in it. Furniture such as event tents, folding tables, and chairs may also be necessary but sometimes these are provided by the event host. Inquire with the organizers.

At local events where there are no tabling opportunities, show up with a pocket full of stickers and put them in conspicuous places or hand them out to the attendees. You don't even have to say a word, just give them something shiny and they will take it home!

FIGURE 14-2: WAYO tabling setup at the Park Avenue Summer Art Fest event in Rochester, New York.

© Max Kelley.

Hosting station spectacles

Get your station out into the community and make some noise! Restaurants and bars are often willing to have station talent set up their turntables in-house for a station DJ showcase. Be sure to bring plenty of swag and signage so everyone knows who you are. These venues will often provide nonprofit organizations with a percentage kickback from the evening's sales, which is fine so long as you are upholding FCC rules by not endorsing the business over the airwaves or encouraging anyone to make purchases (see Chapter 10 for more on FCC rules). Although you can promote the kickback percentages online, I advise not mentioning them on air as this can be seen as an inducement to buy.

Stations often host their own fundraising events for which they go all out with bands, beverages, activities, and more. Perhaps you can get a local college or government agency to donate the space and you can really throw down! DJs enjoy helping with these types of events, and you may find you have everyone you need — performers, bartenders, bouncers, and more — right there among your staff. Local breweries and restaurants have been known to donate food and beverages in exchange for on-air announcements and sponsorship privileges. Done right, this is a win-win for everyone involved.

In-person events can be a lot of work but they build a lot of memories! They can do a great job at raising money for your station and for building camaraderie among your team. The community always enjoys meeting their favorite DJs in person and for that reason alone, stations should consider hosting at least one major event per year (see Chapter 8 for more detail on hosting events).

Here are a few events that may inspire your own ideas:

- WXOX hosts an annual staff talent showcase called *Voix de Ville* ("Voice of the City"). All of the acts in this multimedia lineup feature station DJs.
- CHIRP Radio (`chirpradio.org`) hosts an annual record fair that attracts exhibitors and collectors to Chicago from across the country.
- KRYZ Radio (`kryzradio.org`) presents an annual poetry reading and performing arts festival that brings the community together for a celebration of local creativity.

Covering local events

Get station correspondents out into your community and let your listeners know what is going on in real life, in real time. Political events, band performances, sports challenges, trivia contests, theatrical performances — live remote broadcasts come in infinite shapes and sizes.

TIP

Start off by broadcasting small, low-pressure events that you do for fun and experience. That way, when a really big, important opportunity arises you will be able to seize it confidently!

When your staff members are representing your station in the field, from reporting on a situation to providing live coverage, you should provide them with press badges that identify them as a member of your reporting team. In addition to your station contact information, this badge should include the name of the staff member, the word "Press" in bold letters, and an expiration date. Providing a website or QR code where interested parties can confirm the affiliation of your staff member may also be useful in heated situations. ART FM also includes the text of the First Amendment on ours, just in case anyone needs a reminder.

Vox pops

TECHNICAL STUFF

What the old timers call "man on the street" reporting is now called a *vox pop*. Vox pop is a derivative of the Latin term *vox populi* and translates as "the voice of the people."

It is always exciting when radio personalities get out of the studio and onto the streets. Sometimes they do so for funny skits, at other times to address important issues. Sometimes these reports may come in live to the studio through a call-in to the studio. On other occasions, these commentaries are recorded on the street and brought to the studio later for airplay.

REMEMBER

Getting community voices on air is what it is all about, and amplifying interesting stories and sounds in the environment should be encouraged among your staff. You can have hand-held recorders such as those made by Zoom or Tascam available for them to check out of the studio when on an assignment. The built-in voice-memo apps in smartphones also sound pretty good these days and are very handy for capturing audio on the go.

Community happenings

Live broadcasting community events is actually easier than you may imagine. Coordinated events of any size will have a *PA system* (which is short for a public address system). It includes speakers and a soundboard that are used to amplify the music and information being presented. If the event performers grant permission, it is easy enough to get a direct feed of audio from the event soundboard with the right cables and a willing sound person.

TECHNICAL STUFF

In most cases, the audio feed is provided through cables with XLR connectors. You should bring a set of your own to any remote event. They must be long enough to reach from the event soundboard to your broadcast setup but short enough to prevent any line loss of sound. The sound will begin to degrade at about 10–15 feet. The longer they are, the more shielding they need. Don't skimp on audio cables! The XLR cables are used as an "out" from the venue's audio console and run into your smaller, traveling soundboard. This soundboard must have a USB "out" through which you can then send the audio to your laptop computer. For reliability, your remote broadcast computer should be plugged directly into the Internet via an ethernet cable. At ART FM, we keep a 500-foot ethernet cable wrapped around a garden hose roller for this exact purpose. Learn more about the logistics of setting up a remote broadcast studio in Chapter 15.

If you are live broadcasting an event, make sure that you connect with all speakers and performers ahead of time. They must agree to uphold FCC law or you cannot broadcast their set. They should also sign performer release forms providing you with the rights to broadcast and possibly rebroadcast their music (see the earlier section, "Securing a release form" for more on this important document). Be sure to get these release forms signed ahead of the event where things will be very chaotic. Event organizers must be careful not to promote the sale of food or beverages while on the FM; if they do so, you must be very careful not to accept any remuneration from those who are making the profit.

Rallies and demonstrations

Spontaneous gatherings or political rallies can create unusual situations. You never know where you will end up or what is going to happen.

Public gatherings can become chaotic at any moment, especially if they are emotionally charged. Make sure your DJs know the dangers and that you are not asking them to risk their own personal safety. You may speak to your attorney about drafting waivers for individuals to attest that they are acting at their own risk rather than at the request of the station. Wearing press badges and bright vests can protect journalists, but sometimes this can put them in harm's way.

At times where hard hats and bulletproof vests are in order, community journalists should think very hard about their decisions and consider other ways to get the story (such as getting permission to simulcast audio from live streamers directly from their websites). It is important that LPFM operators get firsthand information, but no single event is worth risking your life or your ability to cover other events for many years in the future.

Reporting on Emergencies

If you build your community relationships well, it won't take long until your programming becomes a part of the everyday lives of your listeners. They will get to know the personalities of your staff and they will come to trust what they have to say. When a crisis hits, they will turn to you. They will expect your help.

Confusion reigns in times of crisis, so listeners need to have a reliable source of information. Never issue an all-clear until you hear it directly from the local authorities.

Emergencies are monsters that come in all disguises. You must be prepared to respond to them by having a list of community resources on hand. Your list of Public Service Announcements (PSAs) for example, should proactively include information critical for surviving an emergency and for dealing with the repercussions afterwards. You can download the free Red Cross Emergency app for access to real-time information and safety tips via the American Red Cross (redcross.org/apps).

Find your state emergency management agency through an Internet search and locate the county directors page on their website. Reach out to those county directors within your listening area and ask to be put their media list. This way you will get alerts about any local situations that may be unfolding. Do not take no for an answer. Although you are a small station, you have a lot of power. Your listeners deserve the lifesaving information when it is available, the same as listeners of big commercial stations.

Sometimes, the EAS alerts that are automatically triggered by the statewide emergency management agencies are not enough. For instance, WXOX realized during the mass shooting at the First National Bank in Louisville that EAS signals do not currently provide a warning when there is an active shooter in the area. However, it was critical to tell our listeners to stay away from the danger. It is important that you always have a way to break into your broadcast with emergency announcements, even if you are playing a prerecorded program.

Unlike commercial stations which are under a variety of contracts to uphold their regular programming schedule, community radio stations have a flexibility that allows for immediate and concentrated response to crisis situations as they unfold. In no instance should-anyone at your station risk their life for the station or its broadcast. The EAS is set to take over in most critical moments. Everyone at your organization should, however, be dedicated to providing honest, up-to-date information — when it is safe to do so — as emergency situations are unfolding. It is true that your radio station can be the difference between life and death, crisis and calm. You are a superhero. Never underestimate your power.

IN THIS CHAPTER

- **Promoting your station to international audiences**
- **Distributing your programming worldwide**
- **Using on-demand tools to share your programming**
- **Getting your remote broadcasting show on the road**

Chapter 15 Connecting with Listeners Locally and Worldwide

We live in an incredible time in which the magic of the Internet can connect us with people worldwide. This includes listeners to your LPFM radio station!

It's one thing to establish a web stream (check out Chapter 5 for more on this), but quite another to have international audiences tuning in. It's hard to describe the thrill of listeners connecting from distant locations around the world. It feels a little like building a bridge to world peace — and I highly recommend you experience it for yourself!

In this chapter, I provide some tips on bringing your brand and your broadcast to locations all around the globe.

Promoting Your Station Internationally

We are living in new and exciting times. Thanks to social media, everyone is a celebrity and everyone can be an influencer. In the following sections I share some ideas that your station can use to attract listeners worldwide.

Getting your swag on

Stickers, T-shirts, keychains, baseball caps, and other swag are some of the best ways to attract attention to your station. When people wear your gear, they are a walking, talking testimonial for your broadcast. There is not much better than that! You can find out more about creating your own merch in Chapter 8.

We love to see ART FM merch in foreign destinations and encourage listeners to send in shots of themselves wearing it in exotic places: hoodies in Norway, coolers in Colombia, baseball caps at the steps of Mordor! You can sell merch and you can also give it away strategically. When big stars visit your station, yes. Please. Give them some merch and perhaps the next time you see that baseball cap you'll be at a Poison concert and Bret Michaels will be wearing it on stage (stranger things have happened)! When students visit the station give them pens, pencils, and buttons. They'll take whatever you've got!

TIP

If you are mailing merch out in the world, don't forget to calculate shipping costs. You'll also need to think about customs fees if you fulfill international orders. For freebies, consider making flat items such as stickers and magnets because these can be sent for the price of a stamp.

Stickers need to be everywhere. Give handfuls to your DJs — especially before big vacations when they can take them out of town. Send packs of stickers to friends in far-off places. Have them posted everywhere you can. You can do this without vandalizing or breaking any laws by finding places that allow stickering, such as community bulletin boards, sticker-covered bathrooms at rock clubs, and on interactive art pieces where additions to the visual atmosphere are encouraged and welcomed. But remember, you don't want to create a bad reputation for your radio station or a bad situation for yourself. Be respectful and, if in doubt, ask before stickering!

Sewing up the socials

Social media. Okay. You know. Use it! The importance of being active and out there cannot be underestimated (see Chapter 12 for more on social media).

REMEMBER

As a station manager, you must engage with social media or have someone reliable on your staff who will. Having your station participate in global conversation on a regular basis is important. In this way you can share ideas and events happening at your station while staying in touch with interesting events around the world. Staying active online is pretty easy for a radio station because there is always a program happening on your station that you can post about. Often you can simply share something the DJ has already created.

Getting published

Radio Survivor (`radiosurvivor.com`) and *Radio World* (`radioworld.com`) both cover happenings at community radio stations worldwide and are interested to know of your goings-on. You may even decide to contact the publication editors asking to write a guest column about something you have learned or encountered in the development of your LPFM broadcast.

TIP

You can subscribe to the *Radio Survivor* podcast and the *Radio World* print magazine by visiting their websites. Both are free and will help you stay in the loop about Federal Communications Commission (FCC) decisions, station activities, and technological developments.

Connecting at conferences

Connecting with other stations is important, locally, nationally, and worldwide. Community radio stations are rare and powerful entities. We are even more powerful when we unite forces!

A great way to connect is through national conferences. Here are a few to check out:

» **The Grassroots Radio Conference (`virtualgrc.org`).** The GRC is the only conference designed for and powered by community radio stations themselves. The GRC is hosted by a different college or community radio station in a different location each year. I was at the Kansas City GRC hosted by KKFI when I first heard about opportunity to apply for a LPFM license and have gained so much information from this conference ever since.

  In 2020, when WXOX had the chance to host the GRC conference, we were thrilled! The COVID-19 pandemic showed up to throw the whole thing for a loop but, with a little experimentation, our staff was able to pull off it off with virtual sessions on responding to the pandemic, the social justice movement, California wildfires, and more. Because the conference was virtual, we were able to get celebrity speakers to attend and we paired them with community members for interviews and which were experienced live by GRC attendees over Zoom and listeners worldwide on the ART FM broadcast. In Figure 15-1 you can see Chuck D of Public Enemy in conversation with WXOX DJ Tia Marie. We were amazed at how well the conference translated to the virtual environment. However, there is nothing like seeing friends in person and we look forward to the next in-person gathering.

- **The College Broadcasters, Inc. (askcbi.org).** The CBI is a great place for college stations to connect with one another and get valuable tips from the pros. Each year, the CBI conference is hosted in a different location and always offers an impressive array of speakers.
- **The College Radio Society (collegeradio.org).** This society celebrates the importance of college radio, manages the College Radio Fund, and coordinates the annual College Radio Day events every October.

  There is also a High School Radio Day (highschoolradioday.com) that takes place every April and keeps high school stations connected throughout the year.
- **The National Federation of Community Broadcasters Conference (nfcb.org).** This conference takes place in every summer, offering education and training to community stations within the public media system.
- **The National Association of Broadcasters Conference (nab.org).** This conference happens annually in Las Vegas, and is geared toward large commercial broadcasters. However, there is always something to learn and the trade floors are remarkable spots for exploring the latest radio gear!

Finding resources and support networks is vital, especially when you are in a minority. The important work is never easy, but you will need to find your allies. Women's Audio Mission (womensaudiomission.org) hosts regular workshops and annual gatherings of women in audio engineering (whom they say make up less than 5 percent of the field). The National Association of Black Owned Broadcasters (nabob.org), the National Association of Hispanic Journalists (nahj.org), the Asian American Journalist Association (aaja.org), and the People of Color Caucus within the Grassroots Radio Conference offer education, professional development, mentorships, and networking opportunities.

FIGURE 15-1: Tia Marie of WXOX interviews Chuck D over Zoom during the Grassroots Radio Conference: Virtual Summit 2020.

Syndicating Programs Worldwide

In order to reach more listeners, producers may want to share their programming with other radio stations. This is called *syndication*. Many programs are syndicated through file-sharing networks such as Pacifica's Audioport and PRX. Programs that are syndicated in live time are called *simulcasts*, as stations simultaneously broadcast the event.

Pacifica Affiliate Network

The Pacifica Affiliate Network (`pacificanetwork.org`) is an organization specifically developed to support grassroots community radio stations. There is a membership fee to become an affiliate. In exchange, Pacifica affiliates receive access to many resources for nonprofit stations including connections to the program producer's email list and the monthly station manager's meet-up where community radio leaders share ideas and experiences.

Affiliates also receive access to Pacifica's Audioport (`audioport.org`), a digital content distribution platform. Over this platform, Pacifica provides simulcasts of congressional hearings, presidential elections, and other new events of national importance for their affiliated stations to pick up live and deliver to their listeners in real time.

Through Audioport, stations have online access to a vast library of historical recordings from the Pacifica Radio Archives (see Chapter 17 for more information on this great resource). Available regular programs include Chuck D's *And You Don't Stop* weekly hip-hop program and Amy Goodman's daily *Democracy Now!* newscast.

After you begin producing your own programs at your station, you can share them through Audioport as well. Then, watch them find syndication on Pacifica affiliates worldwide! Community stations with regular volunteer turnover seem hungry for quality content.

Public Radio eXchange

Public Radio eXchange (`PRX.org`) is an independent, nonprofit multimedia platform creating and distributing programs that are heard on hundreds of public radio affiliates nationwide. Recognized PRX programs include *Echoes* by John Diliberto and *The Takeaway* produced by WNYC. As indicated by its name, PRX focuses on public radio programming, although some community stations participate in its programming exchange. The cost to join is based on the annual station revenue and the number of hours you want to download. PRX producers can earn royalties on distributed programming.

Although you may hear the terms used interchangeably, there are key differences that make "public radio" and "community radio" distinct. *Public radio* is a term most consistently applied to National Public Radio (NPR) affiliates. Although they are usually independent nonprofits, these stations pay thousands of dollars each year for access to programming supplied by NPR producers in Washington, D.C. The station may have some local programming in between these national syndicated shows but this is almost always hosted by paid professional staff. Conversely, *community radio* stations usually do not play NPR syndicated shows and instead focus on live, local productions. For the most part, community radio DJs are volunteers. The program hosts have much more creative freedom on a community station than they would on a public station. As a result, the public and community radio stations have very different sounds and a completely different vibe.

BEING HEARD WAY OUT THERE

Ursula Ruedenberg of the Pacifica Affiliates Network shares her advice on syndicated content distribution:

"The day may arrive when you want to be heard nationally, or even globally. You can do this! The technology exists and there are pathways for you to venture out into the larger media arena. Many radio stations have come to welcome distributed content for the diverse and interconnected sound it offers.

Should you syndicate your show? Ask yourself if you are willing to put in the minimum of 30 to 60 extra minutes per show to prepare it for distribution. Are you prepared to promote this show or have someone help you with this? Discuss this with the management at your local station. For station managers: Stay open to suggesting distribution to your radio hosts when you want to show support for good work. Your station builds pride when taking its place at the bigger media table and helps build the field of grassroots media.

If the light is green, here are some guidelines for getting started:

Decide on the length of the show: 57 and 29 minutes are conventional time frames for radio schedules. Also, 5 or 15 minutes for short segments. Keep the length consistent — to the second! Using music at the end is the easiest way to manage this.

If you are using a recording of your local show, develop a strategy for removing the local-only parts. For instance, have hosts do local announcements or topics within a particular window of time (for example, the first 5 minutes) so that part can easily be edited out or replaced by music or features. Be sure to remove forward promotion of local shows, legal station IDs, local announcements and local underwriting, or local fundraising.

Try to remove day and time references. Stations will air the show at a different time and day.

You may say: "This is [name of the show], coming from the studios of [your radio station]." You may not say: "You are listening to [your radio station]" or "You are listening to [name of the show], at [your radio station]." This would be seen as a violation of Federal Communications Commission (FCC) restrictions.

Remember that a relaxed formal tone works great locally but not to the larger audience. Expect to speak with agency and courtesy in distributed content.

Take inventory and confirm your team and basic technology. Do you need an audio editor? Do you have someone to upload the show for distribution?

Check your sound quality. Good sound quality is everything!

Because you will be using the Internet for distribution you will need compressed audio files. Generally, plan for .mp3s with a bit rate no lower than 128 and sample rate of 44,100.

Be legal! Reread the regulations from the FCC and the IRS. Distributing shows with violations causes great anxiety and is the fastest way to widely ruin your reputation.

If you know who is airing your show, you may want to acknowledge them in the show. Offer to pitch for them during their fund drives, and extend other courtesies. Personal relationships count.

Give each episode of your show an evocative title that tells program managers what the show is about. This greatly increases chances they will try you out.

Produce at least three *evergreen* shows (that can be played anytime) as reserve backup. You may also choose to distribute one-off shows, interviews, documentaries, or limited series.

Decide where to distribute the work. Easily usable distribution services include Pacifica Network's `Audioport.org`, `PRX.org`, a public media group, and `Radio4all.net`, from the A-Infos Radio Project. Go to their websites to find out how to include your work. You can also make a website and make your work available there for download. You may also want to make a version to podcast.

(continued)

(continued)

Make your show available regularly on schedule. At first, this will be less than satisfying because you are not likely to have many takers. Radio stations need time to take on your work, but even before they do, they are checking to see if you are reliable.

The best way to promote your show is to call people — whether it is a distribution group or the staff of individual radio stations. Have your elevator speech ready, describing your show's Mission, and have demos easily available.

Be moneywise. In commercial radio, money is earned by selling time for commercials but with noncommercial radio, this is not allowed. Traditionally you are not expected or allowed to gain financially for distributing content, but to serve the public good with your work. There are some avenues to raise money. You can apply for grants or solicit a national underwriter — be sure to know the rules and regulations about how to do this legally.

Finally, explore what else is out there! Get to know others in the distribution world. At the very least, see who else is distributing content at the distribution sites. It's a great, inspiring, and even collaborative world out there, and there is room for you in it! You can be a valuable addition to the big conversation!"

Providing On-Demand Listening

On-demand listening refers to programs that are recorded and available for replay at the whim of the listener. On-demand listening is a great tool for building your global audience, because it allows them to enjoy your programming at times when it is most convenient for them to tune in. For example, ART FM host *The Freedom Hour*, a radio program providing weekly updates on the female-led revolution in Iran. It airs at 6 p.m. Eastern Time which is 1:30 a.m. in Tehran. If it weren't for our on-demand listening options, made available through Spinitron (`spinitron.com/wxox`) our Iranian audience would not have a viable opportunity to engage with this important show.

WARNING

The *Digital Millennium Copyright Act* (DMCA), found online at `copyright.gov/dmca`, establishes a number of limitations on the ways a radio station may share copyrighted material. For instance, a station may not keep an archive of any given program available for online listening for longer than two weeks' time. Likewise, that radio program may not be *scrollable* (the listener cannot be permitted to navigate to a specific moment on the timeline).

The following sections detail some of the different platforms available for making your content available on demand.

Ark and Confessor

Stations may want to keep two weeks of their programming available online but they must remember to remove the material after that time limit expires, as outlined in the previous section. Spinitron's Ark (`spinitron.com`) and Pacifica's Confessor (`https://pacificanetwork.org/pacificainternetpackage`) offer online solutions that help automate this process and provide stations with the means of embedding these archives into their own website.

DJs who do not record their programs live can go back and record them from the Spinitron Ark or Pacifica Confessor database within 14 days from the program being posted. If they wish to share the episodes from there, things get a bit more complicated in terms of music licensing. The statutory licensing expires after these first two weeks. After that, the program producer needs to get specific permission from each of the copyright holders for permission to share the episode. This means reaching out to each artist or their management individually. If there is a lot of music used in the program, this can be a very time-consuming process — especially for any kind of regular show. If the program does not use any copyrighted material, no worries! — it may be shared freely online on a website and as a podcast, so long as the station management agrees. If the station hosts podcasts or archived program episodes, the DJ should respect that and take advantage of these in-house resources rather than marketing them through a third-party platform. Stations will likely want to develop their own policies regarding the distribution of programming that is made in-house (see Chapter 10) for more on station policies).

Mixcloud

Mixcloud (`mixcloud.com`) is a popular site that handles music licensing for DJs so they can freely share their programs beyond the two-week time frame discussed earlier in this chapter. Previously this was a free site but the platform now charges DJs who upload more than ten programs ($15 per month). Mixcloud embeds advertising onto the program pages but not into the audio recording.

DJs who pay the Mixcloud fee should also consider an equal donation to the station that makes the program possible in the first place.

Soundcloud

If you own the copyright to all the material within your radio program, you may publish it to Soundcloud (`soundcloud.com`). This platform can be particularly good for uploading interview segments or original compositions. Many podcast producers work with local musicians who are willing to share the rights to their compositions for trade, recognition, or a reasonable fee. This can be an excellent way to amplify some local talent and add a unique musical aspect to the show.

Stitcher

Stitcher (`stitcher.com`), now owned by Sirius XM, is the largest podcast listening platform in the world. To add programming to Stitcher, create a free account through the Simplecast Creator Connect (`simplecast.com/stitcher`) and upload content from there.

WARNING

To turn your radio program into a podcast for on-demand listening you need to have the rights to use and distribute all of the copyrighted material it contains. You can review some copyright information in Chapter 10 but also get the input of your attorney so you can avoid any possible violations from the beginning.

I recommend the delightful resource *Podcasting For Dummies*, by Tee Morris and Chuck Tomasi (John Wiley & Sons, Inc., 2020).

Rambling Remotes

Embrace the wonders of technology! During the COVID-19 pandemic, we all discovered how to connect over long distances though the safety of our electronic devices. Out of necessity, ART FM learned that we can run our entire radio station far away from the studio itself. DJs can turn their basements, garages, guest rooms, and dining room tables into remote broadcast studios. We can pick up their signal in the studio and send their voices out to the world! Figure 15-2 gives you a peek at some of the WXOX remote studios, noting that some are very complex but others are surprisingly simple.

Going from one in-person studio to over a hundred remote setups was a technological nightmare but, somehow, we did it! The result was that we reinvented ART FM as a decentralized radio station and proved that we could not only survive, but actively thrive, while providing our city with the essential information it needed for such an unprecedented global crisis. Although the DJs of WXOX are thankfully returning to the downtown studio now, the skills we learned during the pandemic continue to help us broadcast from any place in the well-connected world.

Establishing the DJ stream

TIP

Radio stations serious about establishing remote broadcasting capacities for their DJs can refer to the detailed online reference guide posted on the ART FM website: `https://artxfm.com/remotestations`.

FIGURE 15-2: Examples of remote studio setups developed by WXOX DJs during the COVID-19 pandemic. Some of them are very simple but others are quite complicated.

© Sharon M. Scott (Book Author).

To begin the process, the station needs to connect with a stream host provider and establish a secondary audio stream or *DJ stream* (in a same way they establish the station's primary stream as outlined in Chapter 5). After the stream is set up, the station gets a direct URL and access codes from the host provider, and chooses their own unique password.

In the meantime, the DJ acquires audio encoding software (such as Audio Hijack, B.U.T.T., or Rocket Broadcaster). After this is downloaded, the DJ enters the streaming access codes and password provided to them by the station management (see Chapter 5), because these are needed to initiate the remote broadcasting session.

Setting up audio inputs

DJs can keep it simple using their internal computer mic and digital files for remote broadcasting. They can easily add an external USB mic for a more sophisticated sound. Interviews with guests can be done via video conferencing applications such as Zoom and Google Teams, but better sound quality is delivered through programs that prioritize audio quality, such as Cleanfeed (`cleanfeed.net`) and Riverside (`riverside.fm`).

TECHNICAL STUFF

To broadcast a setup that includes multiple audio inputs such as turntables, CD players, MP3 players, musical instruments, and more, a remote DJ needs a soundboard to accept these audio inputs. Figure 15-3 shows a diagram of a remote studio setup with multiple audio inputs that feed into the soundboard.

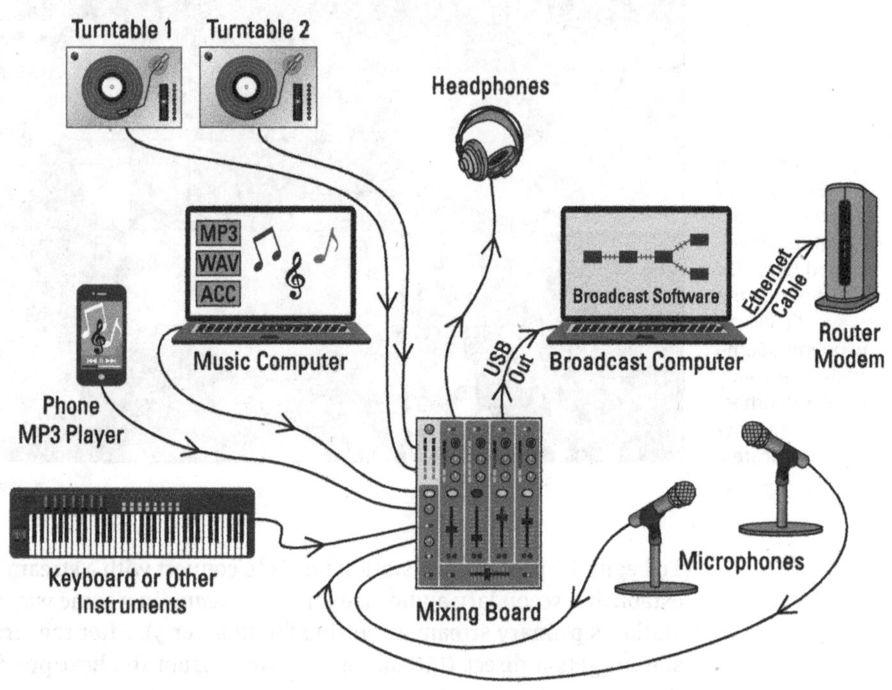

FIGURE 15-3: Author's diagram of a remote studio setup.

Taking flight

The DJ's soundboard should have a USB out that can feed a dedicated broadcast computer. The audio moves through the USB out on the soundboard into the broadcaster's personal computer. Using audio encoding software on the computer, a stream is generated and sent to the online stream host.

After the DJ is connected, the stream is available online through a *direct URL* (a unique web address). The in-house computer at the radio station picks up this URL through a web browser or automation system and feeds the audio to the soundboard. After the input is activated, the audio travels out to FM transmitter and main station stream. Boom! Your remote DJ is on air!

Entering a brave new world!

After your DJs have gained their remote wings, they will be able to broadcast from anywhere in the world with little more than a computer and an Internet connection.

No more excuses for missing their radio show! Whether they are at home sick or traveling to the far corners of the world, DJs can still host their radio program as usual. In fact, the listeners might not even notice a difference — and when they do, I promise they will be intrigued! We've had ART FM DJs broadcast from Los Angeles, New York, Berlin, Tokyo, and Adelaide. We've had them broadcast from a duck pond in Stuttgart, Arkansas, and a campsite at the Clifftop Appalachian String Music Band Festival. Thanks to the wonders of the Internet, the broadcasting possibilities are suddenly limitless. Embrace the technology! Enjoy this new adventure!

6

The Part of Tens

IN THIS PART . . .

Discover ten tricks for defeating radio silence.

Experience ten amazing moments in radio history.

Be inspired by ten Low Power FM veterans who have been through it all before!

IN THIS CHAPTER

» Reacting promptly to on-air problems

» Pulling in resources from outside your radio stations

» Getting new talent on air, quickly!

» Thinking outside the box

Chapter 16
Ten Ways to Fill Dead Air (and Come Alive!)

This is not a dream but it feels like one. Everything is dark but you can feel that you are standing on stage. Suddenly you are blinded by the heat of the spotlight. As things come into focus you realize that they are there — a large invisible audience, silently waiting for you to perform, yet you have no script prepared. This chapter offers my ten top tips for what to do next when you have an attack of *dead air* (radio silence, or whatever the heck you want to call it when suddenly the audio goes quiet and your broadcast is not making a sound!).

Explore Your Music Stacks

This is exactly why you have built a large and expansive music collection within the studio. You find yourself in-studio on many occasions suddenly needing something to play and it's wonderful to have things you can pull off the shelves and throw on the air.

WARNING

The danger of course is breaking FCC laws if you are not 100 percent familiar with the music you're about to play. This is why your in-house albums should be programmed to tell the DJ specifics about the music, such as how it starts and stops (see Chapter 7), but most importantly whether or not it is acceptable for airplay (see Chapter 10).

Stock your station with long-playing instrumental and foreign language records that can be put on air worry-free at any time. Ravi Shankar, Alice Coltrane, and Hawkwind are some of the go-to musicians that DJs know they can rely on in a jam to provide extended mixes that are reliably enjoyable and radio friendly.

Utilize Copyright-Free Material

Thanks to the wonders of the Internet, there is a wealth of audio material available for use on your nonprofit, community radio broadcast. Some material is provided for educational use by its authors with a Creative Commons license and other material has aged out of its federal copyright protection and entered the public domain. When you see either of these phrases, that work is free to use, edit, and redistribute. However, copyright law has changed its definition of public domain several times over the course of the 20th century. Likewise, each country has its own copyright guidelines. Therefore, when dealing with this tricky business it is a good idea to visit `copyright.gov` or consult your attorney to get the full insight.

Generally speaking, in the United States a work is now copyright protected for the life of the author plus 70 years. In the case of a corporate or unknown author, the copyright lasts for the shorter of 96 years from publication or 120 years from creation. Musical compositions produced in the early half of the 20th century have a maximum term protection for 95 years after date of publication. Works created before 1928 are currently in the public domain, and that year increases accordingly every year until it hits 1977. Artists that are interested in sharing their material freely yet still maintaining credit for their work, may get a *Creative Commons* license. This license means different things to different users. When nonprofit organizations see the term Creative Commons or its initials "CC," in most cases they are free to share, copy, and redistribute the material in any medium or format, and adapt, remix, transform, and build upon the material. Attribution for the artist is required. The material may not be used for commercial purposes. Check out `https://creativecommons.org/choose` for the different CC licenses. If you reside outside the United States, please consult your own country's rules.

The first day of every year, Duke School of Law celebrates Public Domain Day (`web.law.duke.edu/cspd/publicdomainday`). On this day, the school lists the works that enter the U.S. public domain and promotes the fact that they are free to copy, share, and build upon. Recent additions to the public domain include Virginia Woolf's novel *To the Lighthouse*, Fritz Lang's film *Metropolis*, and a number of musical compositions from Louis Armstrong. Many broadcasts from the so-called Golden Age of Radio are also becoming available for common use.

A number of sites provide access to royalty free material that can be used on your nonprofit broadcast:

- The Free Music Archive (`freemusicarchive.org/home`) was created by Ken Freedman and a team of colleagues at WFMU as a means for ensuring that community radio stations would be able to share new music with their listening audiences even if online royalty rates became unaffordable. Today the Free Music Archive offers a massive library of songs that are completely royalty free. It's something of a seed library of music, and musicians should be encouraged to sign up for a profile on the Free Music Archive and add some tracks to the mix!
- The Internet Archive (`archive.org`) is a library of digital artifacts that includes digitized books, websites, and audio files. Those that are designated Creative Commons or public domain are free for noncommercial use.
- LibriVox (`librivox.org`) strives for the "acoustical liberation of books in the public domain." Through the LibriVox website, volunteers sign up to read portions of books that are in the public domain. They then upload their recordings and make them available free for the whole world to enjoy!
- C-Span (`c-span.org`) provides live coverage of U.S. government hearings and events. It makes these sessions available for noncommerical use as long as they are credited. Be careful to avoid airing any creative productions by C-Span such as films and documentaries because these require additional permission.
- NASA (`nasa.gov/connect/sounds`) has made thousands of recordings from space missions available online. These files include atmospheric sounds from Mars as well as human voices in outer space and may be downloaded, edited, and freely redistributed. The Chandra X-ray Observatory has a particularly beautiful collection of sonifications on their YouTube channel (`youtube.com/@ChandraXray`). To create these pieces, NASA scientists, led by Kimberly Arcand, assigned musical notes to concentrations of gas and dust in nebulae, black holes, and supernovae. Musicians then performed these pieces to allow more individuals, such as those who are blind, opportunities to interact with outer space.

Amplify Your Voice

Sing! Everyone can, and it is so beautiful! Or, if you're not so sure about your singing range, just talk. Sometimes it can be really enjoyable to hear someone share what is on their mind. Other times it is absolutely grueling to listen to someone ramble on and on, even when they have a lot of well-researched information to share. It is hard to describe what makes the difference but I suppose it just comes down to whether or not you are a good conversationalist. Some of the most well-known radio hosts of our age simply turn on the mic and rant. Millions of people tune them in.

Conduct an Interview

A surefire way to add interest to your airwaves is to interview a guest (see Chapter 14 for some helpful tips on interviews). Everyone has something to say and most people will welcome the chance to speak to your listeners. So, if you are in a jam, look around, find someone to talk to. Perhaps there is a sculptor that lives in your neighborhood who would be interested in dropping by and talking about their craft? Maybe there is a maintenance crew that takes care of your building. Why not bring them on air to talk about their career, secret passageways they've discovered, and any supernatural occurrences they've witnessed in the building? Perhaps one of your colleagues brought their child to work, so why not get their perspective on American politics and current events? Kids tend to be very brave on the microphone and it's good to get them started young.

If there is no one nearby to interview, reach out to a friend or acquaintance through your phone or computer. If you don't have a *telephone-hybrid* (an interface that converts your telephone line into an audio input), plug your mobile device into the soundboard using the headphone jack. This way you can broadcast the feed from communication applications such as Zoom, WhatsApp, Rosebud, and others. Perhaps you can reach out to some artists and small business owners giving them an opportunity to share their latest projects with your listeners. Just be sure to inform, not promote. WBAI radio host Bob Fass filled the weeknight airwaves from midnight to 6 a.m. with call-in conversations about anything and everything on the minds of his listeners. Totally free-form, his show *Radio Unnamable* created some of the most important radio of the 20th century (find out more about Bob Fass in Chapter 17).

Play an Instrument

Time to make some noise! What have you got? A guitar, a keyboard, a noise machine? A triangle, a harmonica, a pair of drumsticks? Can't play any of those? Well, how about a piece of paper, your voice, and a set of drinking glasses? Look around for what is at hand and think about how it can help you fill the airtime. Everyone has music inside of them waiting to make its debut. Even if you aren't a trained musician, you can make improvised noises inside your studio that resemble contemporary directions in composition.

Rally Local Performers

If you don't feel confident making music yourself, perhaps you can rally some local performers. Musicians love coming on air and sharing their songs with the world. If you are well connected within the music community, you should be able to make a few calls and find somebody to venture down to the station and play a set. Actors, likewise, always enjoy a stage. You may be able to feature a play or poetry reading out of a local thespian or literary group. Just be careful that you get all the proper performer releases and have copyright permission for all the material you intend to broadcast.

Instead of having performers in-studio, you may want to go out to where they are performing. If you know of a cool event happening in town, live remotes are always very fun methods for connecting with your community. As long as the talent and the event organizers are willing, you can likely get a live feed from the venue soundboard to share music, lectures, and other activities with listeners all over the world. This is a fantastic way to entertain your audience with fresh material, build relationships with the local community, and amplify the impact of the cultural event.

During remote broadcasts, you need to be vigilant about upholding Federal Communications Commission (FCC) laws. Before hauling all of your equipment out to the venue, confirm with the performers and presenters that their set will be suitable for broadcast, and make sure that they are willing to sign the station release forms. Learn more about the ins and outs of live remotes in Chapter 15.

Simulcast Another Low Power FM Station

Making friendships with other community radio stations goes with the territory, and you'll be able to partner with them for a variety of broadcasting events. If the time comes that your station is unable to generate programming maybe you can simulcast the broadcast of another LPFM station.

Although the FCC prohibits Low Power FM stations from simulcasting the broadcast of full-power FM radio stations, it does permit simulcasting of other LPFM radio stations. This is easy enough to do if the station streams their programming online. You will simply need to pick it up through your station browser and play it out through the soundboard. You may have them announce your legal ID along with theirs or insert it yourself every hour.

Simulcasting another LPFM station is a good option for organizations that are going through transitional phases or need time to train and recruit volunteer DJs. Community radio stations need to spread their wings, and it would be much better to simulcast an existing LPFM for a while than it would be to turn your frequency back into the FCC. Depending on your agreement with the station, you may even be able to add underwriting on your end to earn money for your station as it gets things rolling.

Place a Mic Outside Your Window

Depending on your studio location, place a microphone outside your window and simply let it roll. In the country, your listeners can enjoy the songs of birds, crickets, or a babbling stream. In the city, you may capture the sound of traffic or a passing train. If you are in a populated area, make sure that it is not an area where expletives are in regular use! This option is even nicer if you're near a park or a jazz club where music is often floating in the air.

Find a Numbers Station

A *numbers station* broadcasts human-sounding voices reading a series of numbers. Day in and day out, numbers stations just read out numbers. Believe it or not, listening to numbers stations can be a fascinating activity. Each has its own character, attention signals, and end of transmission alters. Some even play clips of music or their own unusual sounds. So wild!

Numbers stations exist in many languages and seem to be coming from locations all over the world. Some of these stations have been reading out numbers since World War II, still very little is known about them. Some radio enthusiasts are obsessive about discovering, identifying, and locating numbers stations. They will give them names that correlate with particular elements of the broadcast. "Yosemite Sam" is a station that often uses samples of the Warner Brothers cartoon character; "The Lincolnshire Poacher," for instance, is the name of a numbers station that starts each transmission with a song of the same name; "Russian Man" is simply named after the voice reading the numbers. The website `numbers-stations.com` keeps information and sample recordings from these mysterious broadcasters. They are cataloged using a system developed by the ENIGMA group in which one letter indicating the language spoken — such as E for English, G for German, M for Morse Code — is followed by a specific number designation. The understanding is that the intended recipients of these messages (whoever they may be) use a one-time decoding key that correlates numbers to letters and spells out messages. This sort of code is thought to be uncrackable.

TECHNICAL STUFF

You will need a shortwave radio to find a numbers station. Antique models can often be found in thrift stores; modern versions can be purchased online. Unlike FM transmission, shortwave signals travel around the world. With your new receiver you will pick up a variety of unusual international sounds. As you are dialing around, it won't take long for you to pick up a numbers station.

Because they are ongoing and ever changing, setting your shortwave radio up to an in-studio microphone can be an interesting way to fill your silent airwaves. It is doubtful that anyone from these secret stations will step up to claim a copyright violation, and if they do, at least you have helped solve one of the greatest mysteries in the modern world! The thing is, however, because it is likely that these transmissions are carrying coded messages, you really have no idea what you are broadcasting. But unless they have the decoding key, neither will any of your listeners!

Get Experimental

There are so many ways to fill the radio dial and 99 percent of them are better than what is on there right now. When there is downtime on your schedule, don't be afraid to use it a little differently.

TIP

In addition to investigating your own ideas, invite other artists to occupy your airwaves. Wave Farm (`wavefarm.org`), WGXC 90.7 FM Arca in upstate New York, for instance, commissions audio artists to build installations at their art installation park that generates 24-hour FM transmissions and live-streams responding to traffic, weather, and wildlife. Wave Farm is one of the few cultural institutions providing grants and residencies for artists working with broadcast transmission.

REMEMBER

Because so few artists have had an opportunity to throw streaks of sound at this audio canvas, that is exactly what I hope you will do! Remember the FCC rules but push open the boundaries and make way for new directions. Bring fresh ideas and content to the FM dial and you will realize that radio is not dying at all —rather it is singing as it stretches its wings and embraces the chance to fly!

IN THIS CHAPTER

» Celebrating the pioneers of radio

» Revolutionizing the concept of broadcasting

» Remembering the moment when radio changed the world

Chapter 17

Ten Epic Radio Broadcasts

The history of radio is full of miraculous broadcasts that have moved communities to epic feats of heroism. It also contains frightening moments that have brought listeners into states of collective panic. Once you get ahold of a radio broadcast facility, you must always remain aware of how your transmission is affecting those who are tuning in.

In this chapter, you will discover ten epic radio broadcasts that — for better or worse — commanded the attention of their listeners and discovered new powers of the wireless medium. May they serve as some inspiration to you.

The Birth of Radio (March 1, 1893)

The invention of radio has been hotly contested — and mired in controversy at the U.S. Patent Office — but most historians agree that Nikola Tesla made the first public demonstration of a radio broadcast in St. Louis in 1893 at a meeting of the National Electric Light Association. It was a follow-up to a presentation he made to a private group of scientists at the Franklin Institute in Philadelphia just a few days earlier. Tesla conducted the demonstrations believing the patents he held would protect his ideas. Initially he was right. When competitor Guglielmo

Marconi submitted his own plans for wireless transmission to the U.S. Patent Office, they were rejected on account of Tesla's preexisting patents. However, in 1904, when Marconi's company acquired big-name backers such as Thomas Edison and Andrew Carnegie, the Patent Office suddenly had a change of heart and Marconi was granted his patent. A year later he won the Nobel Prize in Physics for his work on wireless telegraphy and radio.

Meanwhile, Tesla's laboratory was destroyed in a fire and he lost most of his life's work. After great misfortune and little credit for his inventions, Tesla died near poverty in 1943. Following his death, the Patent Office reversed their decision and stripped Marconi of his radio patent stating his ideas had been previously registered by other inventors, but they did not mention any specific names.

The World's First DJ (April 1, 1914)

It must have felt phantasmic hearing the disembodied voices of Sybil Herrold and her laughing children as they traveled across the airwaves of San Jose out to listeners tuned in with their handmade devices at home.

The Little Ham Program took flight every Wednesday at 8 p.m., playing requests and the latest musical releases that Mrs. Herrold, a teacher of wireless communications, had borrowed from a local furniture store. Unique at a time when most wireless operators were sending spoken messages across the airwaves, Sybil's selections were entertaining audiences through the experimental radio station belonging to the Herrold College of Wireless and Engineering. Beginning in 1914, her weekly music program, thought to be the first of its kind in the world, was a means for attracting attention to the school that she operated with her husband Charles "Doc" Herrold.

During the *Little Ham Program*, Sybil would refer to her listeners, which included many of her technology students, as her "Little Hams." The Herrold's own young children would often join her on air to scream, giggle, and make funny noises. Doc Herrold was behind the scenes all the while, listening, fine-tuning, and trying to make the transmission of music, noisy infants, and Sybil's gentle voice sound as clear as possible for those receiving at home.

Many of Sybil Herrold's techniques are still used in broadcasting today. In addition to playing requests, she conducted on-air contests through which she collected important information about the location of her listeners and the distance of the school's wireless signal. Eventually, "the experimental station out of San Jose" was licensed as KQW and then transitioned into KCBS at the University of California at San Jose. Now an all-talk radio station held by the megacorporation Audacy, Inc., KCBS is arguably the world's first radio station that remains on air today.

TIP

You can learn more about the Herrolds and their College of Wireless and Engineering on the History San José website (historysanjose.org/charles-and-sybil-herrold). In 1959, Professor Gordon Greb of San Jose State University recorded an interview with Sybil Herrold, which is also available online (audacy.com/media/audio-channel/sybil-herrold-1959-interview-with-gordon-greb). What a treat to hear her voice!

The First Live Broadcast of Presidential Election Results (November 2, 1920)

In October 1920, KDKA was the first radio station to be licensed by the U.S. government. They were allowed 100 watts and broadcast from the top of the Westinghouse Electric and Manufacturing Company in downtown Pittsburg, and engineer Frank Conrad provided the radio expertise to bring it to life. When KDKA aired live results from the 1920 presidential election between Warren Harding and James Cox, it would revolutionize the way news was delivered. On a live program running from the time the polls closed at 6 p.m. until the race was called at noon the next day, host Lee Rosenburg gave commentary and kept listeners locked to their radio anxious for the next update.

In an instant, news delivery was revolutionized. Now, instead of waiting for each morning's newspaper, Americans could get their information immediately as it unfolded. The listeners enjoyed the play-by-play as the election results came in. It kept families glued to their radio sets and it didn't take long before KDKA applied this technology to sporting events to create a radio boom that caught on like wildfire across the United States and has been raging actively ever since.

The Broadcast of War of the Worlds (October 30, 1938)

"As thirst can create mirages in the desert, fear can conjure up sensory images that have no objective reality," Howard Koch stated in *The Panic Broadcast* (Avon, 1970). In this book, he recalled his experience translating the H.G. Wells story *The War of the Worlds* into a radio script for the incorrigible Orson Welles and his legendary Mercury Theater. This radio presentation revealed as much about the mentality of the general population as it did of the power of broadcast radio — and the fiery combustion that results when the two are combined.

Koch developed the script as fictional live musical performance interrupted by a series of short, and increasingly frequent, news bulletins about the unusual occurrences happening on the planet Mars. The interrupting bulletins became more and more frequent until the Martians landed on Earth and began aggressively destroying cities. The effect of the "broadcast within a broadcast" made the imaginary invasion incredibly real to American listeners perched on the brink of another world war. While reports of suicides as a result of this broadcast are unconfirmed, extensive documentation records people running in circles outside of their houses, disregarding all the rules of traffic, and frantically heading towards some unknown destination. Koch continued, "Since by this time my fictional Martians were landing all over the country, it is difficult to understand the advantage of flight, but in a panic situation adrenaline, not reason, becomes the motor power governing behavior."

REMEMBER

There has yet to be another broadcast that has caused such a public commotion, primarily because the U.S. government enacted new laws in the wake of the *War of the Worlds* to prevent another episode of nationwide chaos. The Federal Communications Commission (FCC) now expressly prohibits stations from reporting news that is false and radio stations are forbidden from playing such hoaxes on the audience.

Every October 30, however, it is tradition for radio stations to replay the Mercury Theater production of *War of the Worlds*. It was released on vinyl and you can often find used copies very inexpensively. In light of the FCC laws, it is permissible to air the performance today because it is quite dated and contemporary audiences will not imagine that the news broadcast is real. Yet it is a fascinating listen, year after year, as the ultimate example of radio's exceptional power.

The Launch of KPFA and the Pacifica Network (April 15, 1949)

Lewis Hill was a conscientious objector during World War II. He and his wife Joy worked in radio in Washington, D.C., but feeling prevented from sharing their own views with their listeners, they packed their bags and moved to Berkeley to start their own broadcast station. Hill imagined a place where artists and thinkers could have free use of the airwaves, a place where they would be attracted to work: "America is well supplied with remarkably talented writers, musicians, philosophers, and scientists whose work will survive for some centuries. Such people have no relation whatever to our greatest communication medium."

As much as he loved radio, Hill believed that the medium was being held captive by corporate sponsorships. In his view, stations were afraid of risky programming because they were afraid of losing contracts. The concept of listener-supported stations, Hill believed, would liberate radio programmers from these fears and allow them to provide an open, creative broadcast space attractive to artists and thinkers. In 1946 he established the Pacifica Foundation to do just that! Through the development of his listener-supported initiatives, Hill created a new model for radio stations that would serve communities, not commercial sponsors.

On April 15, 1949, Pacifica's first radio station, KPFA, took to the airwaves in Berkeley, California, and it has been providing alternative news and views ever since. The Pacifica Archives contain some of the richest human rights–related recordings in the world, including broadcasts from Dr. Martin Luther King, Jr., Cesar Chavez, John Trudell, Gloria Steinem, Dolores Huerta, James Baldwin, and more.

The Pacifica Network (`pacificanetwork.org`) now owns stations in five major cities: KPFA in Berkeley, KFPK in Los Angeles, KPFT in Houston, WBAI in New York, and WPFW in Washington, D.C., and over 200 independent stations are members of the Pacifica Affiliates Network. Pacifica, the pacifist network, remains a listener-supported oasis of creativity and free speech on the FM airwaves.

The Launch of First Black Owned and Operated Station in the U.S. (October 3, 1949)

"Dr. King had his office right under the radio station . . . I'd let the mic out the window to the first floor, and Dr. King would bring the mic in the window and make a speech" reads a quote from a WERD "Jockey Jack" Gibson on the National Historical Marker outside the Prince Hall Masonic Lodge on Auburn Avenue in Atlanta, Georgia. The building was once home to both the Southern Christian Leadership Conference (SCLC) and the first Black-owned, Black-operated radio station in the United States.

WERD got its start when Jesse B. Blayton, bank president and Atlanta University professor, purchased the 860 AM frequency in 1949. He appointed his son Jesse Blanton, Jr., as the general manager of the 1,000-watt station, who in turn brought aboard a host of DJ talent that would soon make WERD one of the most popular stations in the Atlanta area. Since it was impossible to segregate the airwaves, white teenagers began enjoying the Black music and culture they didn't have access to before. Through songs that dealt with emotions such as love and jealousy, the shared humanity came shining through. The music of WERD eroded the lines of segregation and embodied Blayton's vision of uniting the races through a radio broadcast. When WERD began airing lectures from Dr. King and weekly simulcasts of his sermons from inside the nearby Ebenezer Baptist Church, the station took a major role in uniting, amplifying, and inspiring the leaders of the American Civil Rights Movement.

Today, the Atlanta Fulton County Library holds a collection of audio archives from WERD (`fulcolibrary.org`). A permanent exhibition celebrating WERD, and displaying some of the many vinyl records from its collection, operates within the Madame CJ Walker Museum (`madamecjwalkermuseum.com`), the building where the station once had a studio.

Bob Fass's Fly-In (February 11, 1967)

Radio Unnamable was a free-form radio program hosted by Bob Fass on Pacifica's WBAI from 1961 until his death in 2021. In the program's prime, Bob was on air from midnight to 6 a.m. answering phone calls, hosting guests, playing music, and offering social commentary. Through his program Bob Fass became the late-night voice of the counter-culture criticizing the draft, chastising politicians, and questioning the Vietnam War. Popular musicians such as Bob Dylan, Richie Havens, and Frank Zappa joined him to provide regular in-studio performances.

Inspired by the architecture of the arrivals building at the John F. Kennedy Airport in New York City, Bob Fass invited his listeners — or *cabal,* as he called them — to join him there for a party. This *Fly-In* had no purpose other than connecting the program listeners in person and welcoming unsuspecting travelers as they landed. Over 3,000 revelers showed up! They brought gifts, musical instruments, and of course, radios! For one night, the arrivals building became a party like no airport had ever expected!

REMEMBER

While the *Fly-In* was a fun nonsensical event, it proved what Fass had long suspected, that there was an invisible tribe listening to his programs that could be motivated for more important purposes. He soon hosted a *Sweep-In* in the Lower East Village where listeners spent the day cleaning neighborhood streets, a *Milk-In* which collected supplies for the Visiting Mothers Daycare that was in financial crisis, and a *Park-In* where attendees flew kites emblazoned with the word "Love" high above Manhattan's Central Park. Fass collaborated with regular guest Abbie Hoffman to stage the first ever *Yip-In* at Grand Central Station. Thousands showed up for this peaceful event but when the hands from the Grand Central clock were removed by young activists "in protest of time" and the police responded violently, *Radio Unnamable* brought the brutal action over the airwaves live to listeners across the city.

On a rare adventure outside of New York City, Bob Fass flew to Chicago to be a part of the Democratic National Convention (DNC) protests in 1968. In the months that followed, he broadcast post-courtroom reports from his colleagues who had been arrested (many of whom would later say the entire idea was formulated in late-night conversations held on Bob's show).

Before his death in April 2021, Columbia University acquired 10,000 hours of recordings from Bob Fass's collection and is in the process of digitizing them so that future generations of cabal have the opportunity to tune into *Radio Unnamable* and understand the influence this free-form program had on American society.

The Launch of Radio Free Alcatraz (December 22, 1969)

"Good evening. This is Radio Alcatraz. We are welcoming you to Indian Land Alcatraz on behalf of Indians of All Tribes." The voice of John Trudell rang across the Pacifica Network live from the main cellblock of America's most notorious maximum-security prison. Santee Sioux John Trudell provided a voice for a group of Native Americans who once again inhabited Alcatraz Island. The occupiers demanded, by terms of the Treaty of Fort Laramie, signed by the Lakota Sioux and the U.S. government in 1868, that Alcatraz Island be returned to the native people now that it was no longer in public use. An organization called Indians of All Tribes (IAT) had plans for building a cultural center on the island where the languages, skills, and traditions of native cultures could be preserved and further developed. In support of this idea, 150 individuals spent 17 months encamped on the island, demanding that the U.S. government live up to its obligations — not only those of Fort Laramie but of all the treaties it had signed with American tribes.

Each weeknight at 7:15 p.m. Pacific Time, Radio Free Alcatraz took to the airwaves and Trudell provided a 15-minute update from the island live on the Pacifica Network. Run on donated equipment and electricity generated on site, it brought the voices of the occupying activists to listeners across Berkeley, Los Angeles, and New York City. Trudell's calm voice explained the background of the situation and helped listeners at home understand why IAT felt the occupation was necessary. While the IAT did not achieve their ultimate goals of reclaiming the island, the occupation of Alcatraz brought attention to the rights of Native Americans and helped push through a number of important U.S. laws protecting their lands and their ability to self-govern.

Thirty-nine episodes of John Trudell's *Radio Free Alcatraz* are available through the Pacifica Archives (`pacificaradioarchives.org`).

The Reporting of Hurricane Katrina and Dome City Radio (August 2005)

There is nothing like an emergency situation to reinforce the importance of FM radio. In 2005 when the intense wind, rain, and hail from Hurricane Katrina pummeled southern Gulf states, one radio station stayed on air at Ground Zero for the entirety of the storm. WQRZ-LP (`wqrz.org`) is a station held by the Hancock County Amateur Radio Association serving Bay Saint Louis, Mississippi, managed by volunteer Brice Phillips. While the area endured wind speeds up to 140 mph, Phillips and his team climbed the tower to move their antenna to three different positions to keep their broadcast live. WQRZ worked alongside the Federal Emergency Management Agency (FEMA) and received special temporary authority from the FCC to increase power from 100 to 4,000 watts during the emergency. FEMA distributed thousands of FM radios to community members so they could receive emergency information from the young but resilient station. When federal agents tried to take full control of the WQRZ emergency broadcast, Phillips refused. He showed them his license and stated that his organization had been authorized by the FCC to maintain the signal and that they were legally obligated to do so. And so they did.

Once the storm subsided, the trouble did not. Flooding in New Orleans forced thousands of people to flee their homes and take shelter in stadiums as far away as the Houston Astrodome. More than 300 miles from home, families were separated from one another with nothing more than the clothes on their back, and there was very little information about what was happening back in the troubled Crescent City. Thankfully, Jim Ellinger had a solution. He convinced the FCC to provide his nonprofit organization, Austin Airwaves (`http://austinairwaves.blogspot.com`), with a temporary permit to broadcast with 30 watts of power on a station nicknamed "Dome City Radio." Austin Airwaves distributed more than 10,000 radios (with batteries and headphones) to the survivors stranded at the Astrodome. Then, from a trailer in the parking lot, volunteer announcers provided news, information, and family relocation services. According to the *L.A. Times*, this was the first documented radio station licensed specifically as a service for victims of an American natural disaster.

The Power of International Human Rights Simulcasts (2020 and Beyond)

Eight minutes and 46 seconds of silence. Station managers nationwide held their breath, in unison, commemorating the life of George Floyd who was killed by Minneapolis police on May 25, 2020. This national simulcast of silence that aired on June 2, 2020, was powerful for those behind the scenes. Those hearing silence on the radio, however, simply changed the channel unaware that something important was going on. In the end, this broadcast failed so magnificently that it reinforced the resolve to get it right. The next time it came time to do a nationwide radio tribute, the participating community stations knew they needed to broadcast something that could really capture the ears of the listeners, cut right to their heart, and jolt their consciousness into action.

On August 13, 2020, at 2 p.m. Eastern Time, over 80 grassroots radio stations worldwide united again, this time to commemorate the life of Breonna Taylor by playing one of her favorite songs in unison: *Everything* by Mary J. Blige. This event, introduced by Breonna's aunt Bianca Austin, and followed by a reading of names, marked the five-month anniversary of Ms Taylor's murder by Louisville Metro Police. It was such a powerful moment, listening to this song, knowing it was going out across the country and being heard by so many people all at once. Together our spines were tingling, eyes were watering, and hearts were crying.

REMEMBER

American grassroots stations have since organized several other unified broadcasts for human rights, including solidarity for the people of Ukraine and the protesters in Iran. It is in these moments that you realize radio remains a massively influential force. One station acting alone is powerful. A network of stations broadcasting in unison is a reminder that when we come together we, the people, have the bandwidth to change the world.

» **Sharing advice for recruiting and managing your team**

» **Finding inspiration for building your LPFM station**

Chapter 18 Ten Inspiring Insights from Low Power FM Station Founders

This chapter introduces you to ten founders of Low Power FM radio stations who are leading their organizations, uplifting their communities, and empowering their neighbors with the tools of radio broadcasting. This is a new generation of community radio heroes! They have been through it all — each in different ways — and every one of them has so much to share. Treat this entire chapter as one big Tip icon from people truly in the know!

REMEMBER

While space is limited in this book, please know that there is a community ready to help you make your station happen! You can do it and you don't have to go it alone. We are here for you! We believe in you! And we will share everything we've got to help you bring your wildest radio dreams to life!

LaGanzie Kale, KLEK Jonesboro

First of all, forget what you thought you knew about radio. This is an adventure like no other. It can take you to places you've never been.

I'd like to share with you the words of my radio mentor, "Don't get caught up in that you've only got 100 watts. You got an FM radio station. Those don't just get handed out." These are words I've carried with me and tried my best to live by.

I've carried these words into the halls of state and national government. I've carried them with me into schools bringing smiles to kids' faces. I've carried them into to the governor's mansion as one of just six individuals to win the Arkansas Community Service Award. I've carried these words as I've spoken with local bus drivers and to candidates for president of the United States.

I've carried these words as I stood alone against a wall of heavily armed officers in full SWAT gear determined to clear a crowd of protesters deeply hurt by the killing of George Floyd, not sure if that would be my last day on earth. I've carried them as I was told KLEK was the reason a person didn't commit suicide. If you take nothing else away from my words, believe that you can make an impact with a LPFM station.

Expect to work harder than you ever have in your entire life. Your success will depend on what you are willing to put into your station! I cannot stress this one enough — if you are not willing to invest in your station don't bother starting it. You will likely be your first donor; if you're unwilling to put your own money into your station, why should anyone else do so?

Nonprofit does not mean "No Money." Bill collectors will come after you just as fast as if you had a for-profit organization. Nonprofit also doesn't mean that you have to struggle or have a mindset that you have to be poor. St. Jude's Children's Hospital is one of the biggest nonprofits in the world, costing over $2 million per day to operate. If Danny Thomas had a poor mindset when he started it, St. Jude's wouldn't be where they are today.

KLEK 102.5 FM is the first Black-owned and -operated radio station in Jonesboro, Arkansas. You can find them online at `www.klekfm.org`.

Julianna Ross, KMGP Seattle

If you are applying for a LPFM radio station, I'm guessing you're a person who does not shy away from a challenge. That's good! Be persistent and build momentum as you go. My top advice for anyone applying for a LPFM station is to be super clear about the reason you are starting a station and keep that in the front of all you do. A solid vision articulated with passion engages and attracts supporters and will be a life raft for you on the days you may hear "no" too often or when the challenges seem too numerous.

As a female with no previous radio experience, I sensed a lot of amusement and dismissal from many people when I told them I was starting a radio station. It's a pretty special thing to do and most people have never entertained the idea. If you've evaluated the situation and potential and are still moving forward, don't let what anybody says or other people's doubts deter you. One of the best parts of volunteering is that you get to use your innate talents and skills in unique ways and you will likely find surprising capabilities within you and your team.

Two hurdles specific to starting KMGP 101.1 was that we had to work through issues with three bureaucracies: Seattle Parks and Recreation, the National Park Service, and the University of Washington. Fortunately, existing relationships and negotiations that kept public benefit as the focus enabled us to move forward and with their assistance, KMGP launched as one of the only stations in the country broadcasting from a public park. Radio is a creative force — use your existing relationships to help build new ones.

I am grateful to the many veterans of the LPFM community who so generously shared their time, offering support and critical information. Without the generosity of people who believed in what we were doing, we could not have gotten on the air. I could give advice and share stories all day, but the key here is to take advantage of the opportunities throughout this process to learn more and make new connections. You never know who is willing and able to help problem-solve and find answers. It's also a good idea to think entrepreneurially or have people who do onboard.

Even six years in, I am still thrilled every day that there is a LPFM radio station in Warren G. Magnuson Park. Although still young in the scheme of things, it is accomplishing its mission to build community and give people a powerful creative outlet for their gifts and energy. The station's audience and diversity of broadcasts continues to grow. The kind comments and little love notes we receive from listeners is worth gold to me as is the chance to broadcast hyper-local information and share good works. I salute your efforts to democratize the airwaves!

Space 101.1 FM broadcasts to Seattle, Washington from Magnuson Park and reaches the whole world on `space101fm.org`.

Shawn Campbell, WCXP Chicago

Community radio is at its best with a team of dedicated volunteers. If you're thinking about starting a LPFM station, one of your first steps should be to consider how you will recruit and train volunteers to support the operation. Some stations have just a handful of volunteers and others have hundreds. But if you want your outlet to serve your community, ideally there should be a group of local folks involved.

How do you find volunteers? You know your community best! Reach out to like-minded organizations, create posters and flyers and share them around town, post on social media, ask to make a presentation at a town meeting, pitch your story to a local publication — whatever works to reach people who might be interested in getting involved with the type of station you're building.

Right from the start, put together a simple volunteer manual with some basic policies and stick to them (many existing LPFMs will be happy to share their materials to help you get started). That doesn't mean you don't listen to volunteer input or let guidelines evolve over time, but it's far easier to start with a set of baseline expectations and adjust them if they're not working than it is to attempt to impose a set of rules after everyone has been doing things freestyle for a while.

The thing is, since most of your staff will be volunteers, it's crucial that they understand how important it is to be reliable, to turn up when they say they will, and stay as long as they've committed to. They need to understand that the station holds a federal license, and that they have to follow noncommercial broadcast rules without exception. But of course you don't want to present this as all rules and no fun! Volunteering at a LPFM is amazing!

At my station, CHIRP Radio in Chicago, we start new volunteer orientations by saying that we have a great time at CHIRP, that we have a wonderful group of people, and that there's nothing like being on the air. But we also take our work seriously. We're committed to our listeners and our fellow volunteers, and that means everyone involved needs to understand that when you decide to volunteer with the station, you're making a commitment. We expect you to be at your shift, on time, every single time. If you sign up to staff a station booth or table at a show, you have to be there. When you're on the air, you're programming for the listeners, not for yourself.

Do some people choose not to continue when they learn that it's not a free-for-all? Sure, but those people probably wouldn't have been very helpful anyway. Having a set of shared expectations where everyone involved understands that they are part of a team working toward the common goal of building a great community-focused station will lead to a healthier, happier, more smoothly functioning LPFM operation.

CHIRP, The Chicago Independent Radio Project, was an early advocate of the LPFM movement, and currently broadcasts on 107.1 FM and `chirpradio.org`.

Max Kelley, WAYO Rochester

Your people (your volunteers and DJs) are your greatest resource — they should be your biggest supporters and can be your first donors! It's important to maintain great relationships with all of them — without them, you won't have a radio station, just a pile of equipment and an FCC license. Get to know them well, check in with them regularly to make sure they have what they need, that things are going smoothly, and that they feel supported by you and the station!

Time, especially at a volunteer-run station, is so incredibly valuable. Find tools (such as online spreadsheets, calendars with reminder emails, and survey tools) to automate tedious tasks so more of your time can be spent on running the radio station — or better yet, use these tasks as an opportunity to get your volunteers involved with your station more!

Keep meeting minutes, checklists, and notes. It's hard to keep track of where you're going if you don't remember what you've already done, and accessible meeting minutes are a great way to provide transparency to the rest of your station members.

When putting together a fundraiser, simple perks with unique artistic designs work great. But make sure that whatever you do, you factor in shipping costs ahead of time — these can be expensive! Use a tool such as Pirate Ship (`pirateship.com`) or `stamps.com` to access bulk-rate shipping prices and print your own shipping labels.

Things can and will get chaotic and stressful, but when it's all said and done, it's an amazing feeling to turn on a radio at any given hour and hear the unique programming that your station members are providing to the community! It's very much a "pinch yourself" moment, and it makes it all worth it.

WAYO, the station that is "Way Out, Right Here," can be heard on 104.3 FM in Rochester and `wayofm.org` all around the world.

Clark Grant, KBMF Butte

Low Power FM stations demonstrate what is possible with the medium of radio broadcasting. While it's important not to denigrate our colleagues on the dial, we must also acknowledge that commercial radio stations and even NPR stations miss the mark on radio's true potential. Community radio is an oasis on the dial. LPFMs give us a chance to claim a vital public resource, the airwaves, for the betterment of our communities and the world.

In running your broadcast operation, always follow the law. Always be fair to those who volunteer their time to add value to your service. Trust everyone initially, and strive to earn and maintain the mutual trust that binds a community together. But also remember that radio is a privilege, not a right. When you encounter those who erode the cohesion of your organization, have procedures in place to protect the resource you've built. Most of all, let kindness, curiosity, and a commitment to public service guide your decision making.

KBMF, "America's Most Radio," broadcasts on 102.5 FM in Butte, Montana, and online at `butteamericaradio.org`.

Davyne Dial, WPVM Asheville

Low Power FM stations are dependent on volunteers for both management and content. This reality can be viewed as a double-edged sword. The absolute best and the absolute worst people will be attracted to your station. You must learn to weed out the worst as soon as possible because they can make life difficult and are a threat to the station.

You need a station policy that people must read, agree to, and sign, and that policy needs to include that being on the air is a privilege, not a right. A year-long trial period in order to see how a volunteer works out is helpful; that way, you can gracefully remove folks smoothly rather than create drama.

Given the opportunity to be on the air with WPVM, all live DJs and commentators (referred to collectively as *show hosts*) agree to adhere to our guidelines. Failure to do so results in the loss of on-air privileges or other actions as determined by the governing body of the radio station. WPVM reserves the right to move a show time, remove a show, or encore for cause or no cause.

Do a background check if possible. If someone does not have a stable background, they probably cannot do what is required to be a good volunteer producer or station volunteer. These three attributes make up a great volunteer: consistency, knowledge, and the ability to communicate with honesty and humor.

One final piece of advice: "Be honest, be consistent, and do not let your fears stop you." I learned this being a self-employed artisan in New Orleans back in the 1980s, and it has stood the test of time.

WPMV, "The Voice of Asheville," is heard on 103.7 FM in the Blue Ridge Mountains of North Carolina and on `wpvmfm.org` worldwide.

Kara Richardson, WXND Louisville

I've been a DJ and have hosted my own show at an awesome LPFM station for 10 years. I love what community radio does for our city. When finding out that another LPFM radio station in our city was going silent and turning their license into the FCC, some friends and I felt the need to save the frequency. We pulled together a small group of people, a selection of community members who are already doing great things in the area, to form our nonprofit. We only had one meeting before the lockdown of the COVID-19 pandemic began. I knew how important the station would be to our community so we kept going while the city shut down.

Then, something tremendous happened.

The station manager of the former LPFM donated their whole local music library to our organization, allowing us to carry on the legacy of Crescent Hill Radio WCHQ-LP, a radio station that solely played local music. It was an amazing gift to give back to our city!

My family and I drove around the area of potential broadcast and searched high points, buildings, and existing towers. I enjoyed the process of the great search and learning the terrain to place our tower. The Radius Maps app (`apps.apple.com/us/app/radius-maps`) helped me narrow down an area. I learned, in my urban area, there isn't a lot of wave space that doesn't conflict with an adjacent channel.

Do you know who you are reaching with your transmissions? We made proposals to many business owners. We finally found a generous supporter of community radio. With the guidance of great engineers, we raised our tower on the tip of their building just ten days before our silent STA expiration. We started our programming there, preserving our local music and extending community broadcasting. It took a year, all of this during the pandemic, and we still got it done! We got the frequency on air! The preservation of the airwaves is ever important.

You will have help. I'm so thankful for the wonderful guidance and expertise from my mentor who manages a station with the biggest heart! Join the networks of LPFMs and listen to their stations. LPFMs themselves are a community! Reach out to them. They are always sharing their experiences, information, and programming. Managing a LPFM involves being creative, inspiring new ideas, and knowing it can be ever-changing, all while finding funding, programming, and volunteers.

Radio is an important resource to enjoy and can be a tool that unites. Every community has unique characteristics that should be documented and local events that need to be covered. Look to embrace them! Your programming will evolve but you need to start somewhere. So always have good vibe intentions and the community's voice will be heard.

WXND is extending the sound of Louisville's community broadcasters on 100.9 FM. Learn more at `www.xtendsound.org`.

David Beaton, WSLR Sarasota

You may be thinking of the birth of your LPFM but have concerns about perpetuating the growth and energy way down the road. First, everyone needs to be cross-trained on all essential functions including fundraising and bathroom cleaning, top to bottom!

Prepare for natural disasters, accidents, and health emergencies. Documenting what you do via video, audio, photographs, and paperwork is critical to transfer knowledge to the next programmer, manager, volunteer, and generation.

Time continues at an ever-increasing speed, so knowledge is ever-evolving. Only a community can slow down time, so build it piece by piece, hand by hand, and hug by hug. Hold your LPFM like a child and nurture it. And set it free once the ground has been made solid. A board, volunteers, programmers, a leader or leaders, and a space to gather are important. Gather together, for it is a community that builds LPFM and vice versa.

WSLR is part of the Fogartyville Community Media and Arts Center in Sarasota, Florida. Their broadcast can be heard on 96.5 FM and `wslr.org`.

Reverend Joan C. Ross, WNUC Detroit

Let me tell you about our journey to becoming a LPFM in Detroit, Michigan. The process was challenging. My background in radio was listening to the Motown music I had grown up with. But Detroit had been racked with outside reporters telling "our story." They focused on how bad Detroit was. And, of course, they only took pictures of the city's worst decay and disinvested areas. We needed a platform for our own voice, to be expressed in our way. We needed a space where "cultural organizing" could take place in our own expressive ways.

But I needed to figure out where to begin. The first obstacle was the application process. We found help from an organization in California called Color of Change (colorofchange.org). Color of Change is the nation's largest online racial justice organization. They are known for helping people respond effectively to the injustice in the world around us. I certainly had some injustice to talk about, and immediately they helped me. They also found one of the country's best engineers to complete the engineering study to determine if a signal was available.

Our application not only had to be approved by the FCC, but it also had to be forwarded to their Canadian counterpart for approval as well (where our signal was based). Seven months later, we got our construction permit.

I wish I could say the rest was history, but it was far from history; the actual ride was just getting started. I had to get local approval, locate the right place for the signal, and raise money to purchase all the equipment I needed to learn the names of.

Two years later we launched. This process would never have happened if, along the way, I had not met some fantastic people. People who will always be superheroes to me. People that I just want to thank right here and now. People like Brandy Doyle, Michael Brown, Betty McArdle, Ursula Ruedenberg, Radio Goddess Donna B, and especially Todd Fisher Wallin and David Klann. It was these genuine radio people who stayed up with me, carried the weight, climbed the tower, and flipped the switch, brought life to 96.7 LPFM Detroit. So my hope for you is that while you're thinking about applying, some superheroes are being dispatched to you, too.

WNUC, a station with a focus on social justice and community development, can be heard on 96.7 FM in Detroit and online at wnuc.org.

Stella Linder Byrne, KMRD Madrid

Even a very small radio station operates by bringing together broadcast engineering, audio technology, logistics, radio content, and people. We don't get to skip any of these pieces! The part that I was initially most excited about when we started building KMRD-LP was the people, and it remains the driving force behind the work that we do. What I mean by "the people" is the way that a LPFM station can effectively serve a community by engaging deeply in the concept of participatory media and access to the airwaves. For our rural small-town station, one of the things that this means is that we know that the station has to be many different things for many different people: a source of connection among DJs, a source of connection among listeners, an artistic medium, a platform for less-heard voices, a (literal, physical) meeting place, an alternative and meaningful way to feel connected to a geographic space.

We believe strongly in meeting people where they're at, and have learned that this includes understanding that different interpretations of and contributions to our collective shared vision are what makes us effectively serve our rural neighborhood. Being clear about what the collective vision is and operating the station with transparency, consistency, and fairness makes room for a truly diverse and committed group of volunteers, listeners, and supporters.

Building a LPFM station is a creative visionary act. Just as the LPFM service secures space on the airwaves for alternatives to mainstream and commercial radio, our individual projects can secure space for alternative visions of community and collective mobilization. New groups building LPFMs don't have to reinvent the wheel to figure out how to put it all together; you have a wide community of mutual support, a network of knowledge! And on the individual station level, my advice is this: Invent, reinvent, and engage the core values and guiding principles that are at the center of why it's important to you to build a LPFM station.

KMRD, a free-form, commercial-free community radio station, can be heard around the clock on 96.9 FM in Madrid, New Mexico, and on `kmrd.fm` worldwide.

Appendix A

Directory of Low Power FM Radio Stations

Check them out! Tune them in! This appendix connects you with a variety of stations that make the LPFM radio community so intriguing. As you read this book, I encourage you to stream into these stations and pay attention to what they are doing with their frequencies. You can learn so much just by listening to them and by surfing their websites. When you travel, catch as many FM signals as you can and arrange some studio visits.

Each LPFM station is powerful on its own. Together we are Legion.

Station Name	Call Sign	Frequency	City	Website	Format
95.7 The Boss	KOWN	95.7	Omaha	957fmtheboss.com	Adult Urban Contemporary
99.3 FM WBTV-LP	WBTV	99.3	Burlington	993wbtv.org	Free-form
99.1 The Beat	KMGG	99.9	Albuquerque	99thebeatfm.com	UAC, Smooth Jazz, Urban
101 Jamz	WVAI	101.3	Charlottesville	101jamz.com	Hip Hop / R&B / Mainstream

Station Name	Call Sign	Frequency	City	Website	Format
101.9 WKRP	WKRP	101.9	Raleigh	wkrpfm.com	Variety Hits
1013rkd Rockford's Best Mix	WRKD	101.3	Rockford	mix1013wrkd.com	Adult Hits
Amazing 102.5 FM	KMAZ	102.5	Houston	amazing1025fm.org	Talk / Variety
Apple Country Radio	WACF	98.1	Brookfield	applecountryradio.org	Variety
ART FM	WXOX	97.1	Louisville	artxfm.com	Free-form / Arts
Bentonville Radio	KOBV	103.3	Bentonville	bentonvilleradio.com	Community News / Events / Music
Big D Country	WDXD	101.9	Tallahassee	bigdcountry.com	Country / Variety
Bisbee Radio Project	KBRP	96.1	Bisbee	kbrpradio.com	Variety
BSR: Brown Student Radio	WBRU	101.1	Providence	bsrlive.com	College / Free-form
Buckeye Broadcasting	WBPS	101.9	Cambridge	buckeyebroadcasting.com	Oldies
CHIRP Radio	WCXP	107.1	Chicago	chirpradio.org	Independent Music
Dolphin Radio	WXDR	98.9	New Orleans	dolphinradio.org	College / Free-form
Downtown Radio	KTDT	99.1	Tucson	downtownradio.org	Adult Album Alternative
Extreme Independent Radio	WXIR	100.9	Rochester	1009wxir.com	Variety
Forward Radio	WFMP	106.5	Louisville	forwardradio.org	Talk / Music
Free-form Portland	KFFP	90.3	Portland	freeformportland.org	Free-form
Free Vermont Radio	WFVR	96.5	Royalton	wfvr.org	Eclectic — Music / Talk
Frogtown Community Radio	WFNU	94.1	St. Paul	wfnu.org	Variety Music / Local Talk

Station Name	Call Sign	Frequency	City	Website	Format
Harvard Broadcasting Inc.	WHIW	101.3	Harvard	whiwHarvard.com	Free-form
Hollow Earth Radio	KHUH	104.9	Seattle	hollowearthradio.org	Free-form
Illinois Valley Community Radio	KXCJ	105.7	Cave Junction	KXCJ.org	Community-powered radio by, for, and about the Illinois Valley
Jazz ATX	KJZX	89.1	Austin	JazzATX.org	Jazz / Blues / Folk
Joppa Radio	WJOP	96.3	Newburyport	ncmhub.org	Classical / Jazz / Local / Talk
KBFG	KBFG	107.3	Seattle	kbfgseattle.org	Local Music / News / Talk
KBMF — America's Most Radio	KBMF	102.5	Butte	butteamericaradio.org	Free-form
KCIW	KCIW	100.7	Brookings	kciw.org	Eclectic
KDIF 102.9 FM	KDIF	102.9	Phoenix	kdif.org	Bilingual / Variety
KDRU 98.1 — Drury University	KDRU	98.1	Springfield	kdruradio.com	College / Free-form
KFOK: The Voice of The Divide	KFOK	95.1	Georgetown	kfok.org	Variety
KLEK 102.5 FM	KLEK	102.5	Jonesboro	klekfm.org	Urban AC / Gospel / News
KLLG — Willits Hometown Radio	KLLG	97.9	Willits	kllg.org	Community / Free-form
Knowledge, Understanding, & Wisdom	KUAW	98.5	Kansas City	kuaw.org	Entertainment / Community News / Talk
Knoxville Community Radio	WOZO	103.9	Knoxville	wozoradio.com	Open / Free-form
KOWS 92.5 FM	KOWS	92.5	Occidental	kows92-5.org	Community Variety

Station Name	Call Sign	Frequency	City	Website	Format
KPPQ-LP Ventura	KPPQ	104.1	Ventura	capsmedia.org/radio	Community
KPSQ - The Public Square	KPSQ	97.3	Fayetteville	kpsq.org	Free-form
KRWV	KRWV	99.3	Gold Canyon	krwv.org	Classic Hits
KRYZ Radio	KRYZ	98.5	Mariposa	KRYZradio.org	Free-form Music / News
KSFP — San Francisco Public Press	KSFP	102.5	San Francisco	ksfp.fm	News / Talk / Storytelling / Variety
KTQA	KTQA	95.3	Tacoma	ktqa.org	Free-form
KUHS	KUHS	97.9	Hot Springs	kuhsradio.org	Variety
KUOZ — University of the Ozarks	KUOZ	100.5	Clarksville	ozarks.edu/kuoz	College / Free-form
KUYI Hopi Radio	KUYI	88.1 / 89.1	Upper Moencopi	kuyi.net	Native American Public Radio
KVRU	KVRU	105.7	Seattle	kvru.org	Music / Community News
KWSI	KWSI	100.3	Grand Junction	kwsi.org	Progressive Talk / Music
KXSU — Seattle University	KXSU	102.1	Seattle	kxsu.org	College / Free-form
KZAX — Make Shift Radio	KZAX	94.9	Bellingham	kzaxradio.org	Free-form
Lakes Community Broadcasting	WLCB	101.5	Round Lake Heights	akesradio.org	Free-form
Lexington Com-munity Radio / El Pulso Latio	WLXU / WLXL	93.9 / 95.7	Lexington	radiolex.us	Free-form / Bilingual
Long Beach Public Radio	KLBP	99.1	Long Beach	klbp.org	Entertainment / Education / Conversation
Lumpen Radio	WLPM	105.5	Chicago	Lumpenradio.com	Talk / Music Multilingual

Station Name	Call Sign	Frequency	City	Website	Format
Madrid Community Radio	KMRD	96.9	Madrid	kmrd.fm	Community / Free-form / Local
The Mix 94.3	KTWI	94.3	Liberal	ktwi.org	Top 40 / Country
Napa Radio Project	KCMU	103.3	Napa	naparadio.org	Free-form
Ohm Radio	WOHM	96.3	Charleston	ohmradio963.org	Free-form
PhillyCAM Radio	WPPM	106.5	Philadelphia	phillycam.org	Free-form
Poder 97.1 FM	WEPL	97.1	Rochester	ibero.org/poder-2	Music / Talk (Spanish)
Portland Radio Project	KPRP	99.1	Portland	prp.fm	Free-form / Local / Indie
The Public Square	KPSQ	97.3	Fayetteville	kpsq.org	Eclectic Music and Talk
Q — 101.3	WQMR	101.3	Rocky Mount	wqmrfm.com	Classic Rock / Oldies
Radio Tacoma	KTAH	101.9	Puyallup	radiotacoma.org	Free-form
RadiOpio — Pa'ia Youth and Cultural Center	KOPO	88.9	Maui	pyccmaui.org	Youth Radio
Real Oldies 97.9 The WREN	WREN	97.9	Charlottesville	979wren.org	Oldies
Richmond Independent Radio	WRIR	97.3	Richmond	wrir.org	Independent Music / News and Views
Riverwest Radio	WXRW	104.1	Milwaukee	riverwestradio.com	Talk / Radio Theater
RockHits 92.3	WXRK	92.3	Charlottesville	923xrk.org	Active Rock
Rocky Mountain Multicultural Community Radio	KETO	93.9	Denver	ketofm.org	Multicultural (African immigrant and refugee)
Smooth Jazz Utah	KWBR	105.7	St. George	smoothjazzutah.com	Jazz
Space 101.1 FM	KMGP	101.1	Seattle	space101fm.org	Free-form from Magnuson Park

Station Name	Call Sign	Frequency	City	Website	Format
Spark!	WSPJ	103.3	Syracuse	sparksyracuse.org	Eclectic
Surf 97.3 FM	WQFB	97.3	Flagler Beach	flaglerbeachradio.com	Beach and Oldies
Takoma Radio	WOWD	94.3	Takoma Park	takomaradio.org	Free-form
The Voice of Athol-Orange	WVAO	105.9	Athol	aotv13.org	Talk
The VOICE of the Community	KIEZ	106.7	Monroe	KIEZ1067.com	Talk
WAKT	WAKT	106.1	Toledo	toledoradio.org	Music / Talk
WAYO — Way Out, Right Here	WAYO	104.3	Rochester	wayofm.org	Free-form
WBCA — Boston Neighborhood Network	WBCA	102.9	Boston	bnntv.org	Free-form
WCAA-LP 107.3 FM	WCAA	107.3	Albany	grandarts.org/wcaalp1073	Arts / Music / Community Activism
WCFA — Cape May Radio	WCFA	101.5	Cape May	capemayradio.org	Music / Variety / Talk
WDZD — Fun One Radio	WDZD	99.1	Monroe	wdzdfm.org	Oldies / Beach
WELT	WELT	95.7	Fort Wayne	acpl.info/welt	Music / Talk
WERA — Arlington Independent Media	WERA	96.7	Arlington	wera.fm	Community Media / Free-form
WGRN — The Green Renaissance	WGRN	94.1	Columbus	wgrn.org	Environment
WHIV	WHIV	102.3	New Orleans	whivfm.org	Health Information / Music / Social Justice
Wimberley Valley Radio	KWVH	94.3	Wimberley	wimberleyvalleyradio.org	News / Music

Station Name	Call Sign	Frequency	City	Website	Format
WNUC —- Detroit's Community Station	WNUC	96.7	Detroit	wnuc.org	Community Media Arts
WNUW — Newman University	WNUW	98.5	Aston	media.neuman.edu	College / Media Arts
WPPM 106.5 FM	WPPM	106.5	Philadelphia	Phillycam.org	Free-form Music / Talk
WPVM —- The Voice of Asheville	WPVM	103.7	Asheville	wpvmfm.org	Free-form Music
WQRT	WQRT	99.1	Indianapolis	wqrt.org	Art / Culture / Music
WRGV — Georgetown College	WRGV	93.7	Georgetown	georgetowncollege.edu/wrvg	College / Free-form
WSLR	WSLR	96.5	Sarasota	wslr.org	Community Media
WTSQ: The Status Quo	WTSQ	88.1	Charleston	wtsq.org	Variety
WURK Community Radio	WURK	96.3	Tampa	wurkfm.org	News / Talk / R&B / Hip Hop/ Gospel / Jazz
WXNA	WXNA	101.5	Nashville	WXNAfm.org	Free-form
WXRU-LP SMOOTH 107.9 FM	WXRU	107.9	Greenville	wxru1079fm.com	Soul / Blues / Gospel / Jazz
WXTJ: University of Virginia	WXTJ	100.1	Charlottesville	wxtj.fm	College / Free-form
Xerocraft	KMKR	99.9	Tucson	xerocraft.org	Free-form Hackerspace
Xtendsound	WXND	100.9	Louisville	xtendsound.org	Local Music and Broadcasters
ZUMIX Radio	WZMR	94.9	Boston	zumix.org	Youth-Powered Radio

Appendix B Glossary

Broadcasters are well known for their buzzwords and balderdash. The following pages introduce you to some of the language used in the biz so that you can sound like a radio professional right from the start!

1st adjacent channel: The position on the broadcast band directly next to another channel.

2nd adjacent channel: The position on the broadcast band that is two positions away from another channel.

Adds (additions): New songs or albums added to a station's music rotation each week.

Aircheck: A recorded copy of broadcast audio used for demonstration purposes.

AM (Amplitude Modulation): This refers to the way an audio signal is encoded on the carrier frequency. AM was the first form to transmit wireless radio signals. Operates at a frequency range of 535–1705 kHZ on the standard broadcast band.

Archive: A library of recorded audio tracks or radio programs.

ASCAP (American Society of Composers, Authors, and Publishers): A performance rights organization that collects and distributes royalties on behalf of songwriters and composers. Find them online at `ascap.com`.

ATH (Aggregate Tuning Hours): Total number of hours listeners have spent listening to an online radio station. These are necessary for calculating streaming music royalty payments.

Audioport: Program sharing system for community radio stations. Developed and managed by the Pacifica Affiliates Network.

Back announce: Information about a track, such as title and artist name, spoken by a radio presenter after playing a song.

Bed music: Audio track that a radio announcer speaks over (usually instrumental).

BMI (Broadcast Music, Inc.): Performance rights organization that collects and distributes royalties on behalf of songwriters and composers. Find them online at `bmi.com`.

Call sign: A four-letter identifier for radio stations. In the United States, stations east of the Mississippi start with a "W" and stations west of the Mississippi start with a "K." The remaining three letters are wildcards chosen by the frequency holder.

Cans: Headphones.

CDBS (The Consolidated Databases System): Used for filing petitions with the FCC. Found online at `fcc.gov/media/filing-systems-and-databases`.

Co-channel: A radio transmission that is on the same frequency as another.

Compression: A system of sound management that reduces a signal's dynamic range, eliminating the highest and lowest levels.

Control room: An on-air studio.

Copy: A script to be read by a presenter.

CORES (Commission Registration System): FCC website for registering and managing a FCC Registration Number (FRN). Found online at `https://apps.fcc.gov/cores`.

CP (Construction Permit): Issued by the FCC, giving permission to build a broadcast station.

CPB (Corporation for Public Broadcasting): A private, nonprofit corporation created by Congress in the Public Broadcasting Act of 1967. Financed by American taxpayers, the CPB is the largest single source of funding available for public and community broadcasters. Find them online at `cpb.org`.

CRB (Copyright Royalty Board): A three-judge panel that works to determine the rates and royalties for statutory licensing that must be paid to the U.S. Copyright office. Found online at `crb.gov`.

CSRS (Call Sign Reservation System): The FCC's online system for searching and registering station call signs.

Cue: To prepare audio for airplay.

Decibel (dB): Unit for measuring sound.

Demo: A short collection of audio pieces to showcase talents of a performer, DJ, or radio station.

Drive time: Popular radio listening hours when listeners are commuting to and from work. In radio, it's weekdays 7–9 a.m. and 4–6 p.m.

EAS (Emergency Alert System): The national warning system in the United States used by state and local authorities to deliver important emergency information.

ERP (Effective Radiated Power): Power that radiates out of an antenna. Low Power FM stations are permitted 100 watts of ERP at 100 feet HAAT.

ETRS (Emergency Test Reporting System): The FCC's online system for reporting test results of a station's Emergency Alert System (EAS). Found online at `fcc.gov/general/eas-test-reporting-system`.

Evergreen: A recorded program that makes no date or time references. It may be played at any time.

Fader: A sliding controller on a soundboard to regulate audio levels.

Feed: An external audio source sent to a remote location.

Feedback: An uncomfortable squelching noise created when sound from a loudspeaker interacts with a microphone and they begin amplifying and distorting one another's sound.

FCC (Federal Communications Commission): The U.S. government agency assigned with regulating interstate and international communications by radio, television, wire, satellite, and cable in all U.S. states and properties. The agency is overseen by Congress and is responsible for implementing and enforcing America's communications laws and regulations. Found online at `fcc.gov`.

FM (Frequency Modulation): The means by which the audio signal is encoded on the carrier frequency. In the United States, 87.5–108 MHz is used. Only noncommercial educational stations may operate on the 88–92 MHz "reserved" band. The "non-reserved" 92–108 MHz band may be used by both commercial and noncommercial stations.

Foley: An audio artist who makes sound effects for radio productions.

Format: The type of programming a station broadcasts.

FRN (Faculty Registration Number): Used for noncommercial and commercial broadcast filings with the FCC.

FX ("effects"): Sounds used to create audio atmospheres for radio programs.

Gain: Volume.

GRC (Grassroots Radio Conference): An annual gathering of community radio stations. Find them online at `virtualgrc.org`.

HAAT (Height Above Average Terrain): Elevation of a transmission point above the local landscape.

H.A.L.: Nickname for an automated DJ. Based on a A.I. character from Stanley Kubrick's *2001: A Space Odyssey* movie.

Headlines: The top news stories summarized. Usually announced at the top of the hour.

Hook: A catchy riff or repetitive lines in a song that become its most memorable part.

Jingle: A short musical tune that's used as a station identifier.

Kut: A track on an album.

LCRA (*Local Community Radio Act*): An act passed by Congress in 2010 and signed by President Obama in January 2011. It helped grow thousands of Low Power community radio stations across the United States. Read it online at `congress.gov/111/plaws/publ371/PLAW-111publ371.pdf`.

Legal ID: The call sign and city of license that must be announced by every broadcast station in the United States at the top of every hour.

Level: Volume. When the levels are too low, the listener cannot hear; when they are too high, the audio becomes distorted.

Live read: A script read on air by a radio presenter.

Log: A chronological list of items played or announced on air.

LMS (Licensing and Management System): The FCC online system for submitting, managing, and tracking applications to the FCC. Found online at `enterpriseefiling.fcc.gov/dataentry/login.html`.

LPFM: A Low Power FM radio station belonging to a specific class of American broadcast stations, exclusively operated by local nonprofit organizations and governed by the FCC.

Mix: Multiple channels of audio, mixed down into one stream.

Monitor: A speaker. Device for "monitoring" (listening to) audio.

Mono: Left and right channels mixed together to form one channel of audio.

MX group (mutually exclusive group): These groups include FCC broadcast applications that are in competition for the same frequency within a limited geographic area. They are mutually exclusive because if one of these applicants is granted, the others will be excluded.

NAB (National Association of Broadcasters): Trade association and lobby group representing the needs of American radio and television broadcasters. Members are primarily commercial stations. Find them online at `nab.org`.

NCE (NonCommercial Educational) broadcast station: A class of broadcast stations defined by the FCC. Sometimes referred to as non-comms.

NFCB (National Federation of Community Broadcasters): Member-based organization founded in 1975 to provide training, education, and advocacy to community and public radio broadcasters. Found online at `nfcb.org`.

NPR (National Public Radio): American media network based in Washington, D.C., providing syndicated programming and support services to over 1,000 member stations nationwide. Find them online at `nrp.org`.

Pacifica Affiliate Network: A community of independent radio stations that share programming and resources nationwide. This nonprofit is a separate business unit of the Pacifica Foundation. Found online at `pacificanetwork.org`.

Pacifica Foundation: Nonprofit community radio network founded during World War II by peace activist Lewis Hill. KPFA launched in 1949, and Pacifica now operates five full-power radio stations in major U.S. cities. The Pacifica Archives holds a vast library of historical recordings made at these stations from the 1940s to the present. Found online at `pacifica.org`.

PA system (Public Address system): Audio system used to amplify music and address crowds. Includes speakers, soundboard, and accessories.

Programmer: Radio host or DJ.

Promo: On-air announcement promoting a station program or activity.

PROs (Performing Rights Organizations): Organizations such as SESAC, BMI, and ASCAP that represent music performers by collecting and distributing royalties on their work.

Public Radio eXchange (PRX): An independent, nonprofit multimedia platform that creates and distributes programs that are heard on hundreds of public radio affiliates nationwide.

PSA (Public Service Announcement): On-air announcement from a nonprofit organization or government agency providing information for the benefit of the listener and the community.

RDS (Radio Data System) feed: Digital information embedded in a FM frequency providing short messages to RDS-enabled receivers.

Release: A legal agreement signed by a guest before going on air, giving the station permission to broadcast, record, and/or redistribute performances or interviews.

Remote: Live broadcasting of a program or event from a location other than the studio.

RMT (Required Monthly Tests): These locally and federally activated tests ensure that the Emergency Alert System works correctly.

Rotation: Repeating playlist used by some radio stations.

Playlist: A group of songs selected by the DJ or station management for airplay.

Royalties: A sum of money paid to compensate artists, composers, and publishers for use of their creative work.

SBE (Society of Broadcast Engineers): Professional organization that offers workshops, conferences, and legislative representation for broadcast and multimedia technology professionals. Find them online at `sbe.org`.

Segue: Seamless, sometimes unnoticeable, transition from one song to the next.

SESAC (Society for European Stage Authors and Composers): Performing rights organization that now collects royalties for songwriters and composers of all nationalities in the United States. Found online at sesac.com.

Simulcast: A simultaneous broadcast of another station's program or event.

Singleton: The only eligible applicant for an FM frequency.

Soundbite: A short piece of audio.

Sound drop: A soundbite from a celebrity, often used as a station identifier on air between songs.

SoundExchange: Nonprofit organization authorized by the U.S. Government to collect and distribute digital music royalties on behalf of musicians and other performers. Found online at soundexchange.com.

Spinitron: A playlist-management system for noncommercial radio stations. Handles playlist entry, archiving, reporting, and publishing. Found online at spinitron.com.

The stacks: Area of the radio station where records and CDs are held.

Station authorization: Construction permit, license, special temporary authority, or any other authorization issued by the FCC.

STA (Special Temporary Authority): Granted to a station licensee, in writing, by the FCC for operation of a broadcast facility for a limited amount of time at a specified variance from the normal terms of the station authorization or requirements of the FCC rules.

Stereo: Two distinct channels of audio, left and right.

Stereophonic sound: A method of sound production seeking to create a multidimensional audio perspective.

STL (Studio to Transmitter Link): Device used to send the audio from the radio studio to a remote transmitter site.

Syndicated program: A radio program broadcast by more than one station.

Teaser: A short audio piece designed to keep the audience tuned in.

TPO (Total Power Output) of FM transmitter: The total amount of power a transmitter produces at its point of output (antenna).

Transition: The movement from one song or one radio program into another.

VU meter: A tool for measuring sound levels of incoming audio.

WAM (Women's Audio Mission): A nonprofit dedicated to the advancement of women/girls/gender-expansive individuals in music production and the recording arts. Found online at womensaudiomission.org.

Index

A

AAJA (Asian American Journalist Association), 244
access codes, 215
accessories, 107
Accounting For Dummies, 7th Edition (Tracy and Tracy), 152
accredited educational institutions, 23
accuradio, 145
Acostimac, 95
Acoustic America, 95
Ad Council, 191
adapters, 107
adds (additions), 293
Adobe Audition, 106
Adobe Sign, 194
AEQ Broadcast, 98
Aggregate Tuning Hours (ATH), 293
Aiir, 85
aircheck, 293
airtime, equal, 190
albums, 118–120
Alexa, 88
Alice's Restaurant, 187
Amazing 102.5 FM, 286
American Society of Composers, Authors, and Publishers (ASCAP), 144–145, 293
amplifying voice, 259–260
Amplitude Modulation (AM), 293
analog boards, 98
Android Application Development All-in-One For Dummies (Burd and Mueller), 85
annual taxes, filing, 153
antennas, installing, 68–69
AnyDesk, 65
Apple Country Radio, 286
applications
  Applications for a Low Power FM Construction Permit, 20–23
  handling for on-air team, 201
  submitting, 49
apps, developing, 85–88
Arakkis Systems, 98
Arcade Automatic, 192
archiver, 106
archives
  defined, 293
  in-house productions, 125–126
  music licensing for, 172–173
Ark, 249
ART FM/WXOX Louisville, 28, 36, 37, 131, 132, 177, 188–189, 201, 232, 250, 286
Articles of Incorporation, 29
ASCAP (American Society of Composers, Authors, and Publishers), 144–145, 293
Asian American Journalist Association (AAJA), 244
assets, protecting, 33–34
ATH (Aggregate Tuning Hours), 293
Atlanta Fulton County Library, 270
Audacity, 106
audio
  setting up inputs for, 251–252
  sources for, 173
audio console, 98–99
Audio Hijack, 81, 106, 214
audioport
  adding to web streams, 81–83
  defined, 293
Audubon Society, 191
Auralex, 95
Austin Airwaves, 21, 64, 272
Avery, 138

B

back announce, 294
Bandcamp, 122
Barix, 65
Barnes, Tim, 182
Bases Loaded Radio, 192
Be Here, 186
Beaton, David, 282
bed music, 185, 294
Bentonville Radio, 286
Beryl, Creighton, 177
B&H Photo, 97
Big D Country, 286
Bisbee Radio Project, 286
blanket licenses, 144, 172
Blayton, Jesse B., 269–270
BMI (Broadcast Music, Inc.), 144–145, 294
board, 98–99
board of directors, appointing, 24–27
Bolhari, Shirin, 231
bookkeeping, 152–153
branding, for nonprofit organizations, 34–37
breaking events, 189–190
Brickell, Edie, 112
Broadcast Music, Inc. (BMI), 144–145, 294
broadcast of telephone conversations, 168
broadcast permission
  about, 39
  alternative methods for obtaining, 55–57
  applying for FM frequency, 46–49
  finding a frequency, 39–42
  MX situations, 51–55
  towers, 42–46
  winning your frequency, 50–51
Broadcast Supply Warehouse, 97
Broadcast Tool & Die, 79, 124–125
Broadcast Using This Tool (B.U.T.T.), 81
Broadcasters General Store, 97
Brown Broadcast Services, 21, 41, 64
Browning, Beverly A. (author)
  Grant Writing For Dummies, 7th Edition, 136
BSR: Brown Student Radio, 286
BSW, 100
Buckeye Broadcasting, 286
Burd, Barry (author)
  Android Application Development All-in-One For Dummies, 85
business manager, 199
B.U.T.T. (Broadcast Using This Tool), 81
by-laws, drafting, 30
Byrne, David, 146
Byrne, Stella Linder, 283–284

C

cables, computer, 106–107
Call Sign Reservation System (CSRS), 294
Call Sign Search page (website), 62
call signs
  claiming, 62–63
  defined, 294
Campbell, Shawn, 184, 277–278
cans, 294
Capell, Laura (author)
  Event Management For Dummies, 156
capital campaigns, organizing, 128–131
Carlin, George, 166
Carnegie, Andrew, 266
CARS, 137
Cars2Charities, 137
cases, 116
cassette players/tapes, 105, 118
CBI (College Broadcasters Incorporated), 150, 244
CDBS (Consolidated Databases System), 294
CDs/CD players, 104, 117–118
celebrities, interviewing guests for, 227–228
chair, of board, 26
Charity Cars, 137
Charity Navigator, 153
Cheat Sheet (website), 4
chief operator, 199
CHIRP Radio: Chicago Independent Radio Project, 184, 278, 286

Chrome Remote Desktop, 65, 143
CiviCRM, 152
Civil Rights Act of 1964, 207–208
The Clash, 112–113
clean music library, 123
Cleanfeed, 251
Clifford, Clyde (DJ), 95, 185
co-channel, 41, 294
Cognito, 201
College Broadcasters Incorporated (CBI), 150, 244
College Radio Society, 244
Color of Change, 283
Comet CFM9557, 68
Commission Registration System (CORES system), 46–47, 62, 294
Commission's Rule 73.1212, 171
Common Frequency, 21, 41, 68, 160
communication costs, 151
Communications Act of 1934, 9, 163, 171, 188
community access/security, providing, 108
community connections
  about, 225–226
  building, 36
  covering local events, 236–238
  DJ training programs, 232–233
  festivals, 234–235
  hosting open studio events, 226
  hosting station spectacles, 236
  interviewing great guests, 227–230
  parades, 234–235
  power of radio, 230–231
  reporting on community trauma, 230
  reporting on emergencies, 238–239
  veterans' stories, 231–232
  working with licensees in your, 55–56
  youth groups, 231
community giving days, 130–131
community happenings, 237–238
Community Media Assistance Project, 21
Community Music, 122
community trauma, reporting on, 230
compression, 294
computers/computer cables, 105–107
Comrex, 65
condenser mics, 99
conferences, for promoting internationally, 243–244
conflict of interest policy, 24
Conrad, Frank, 267
Consolidated Databases System (CDBS), 294
Construction Permit (CP), 61, 294
content regulations, 165–169
continuing education, funding, 149–150
control room, 294
copy, 294
Copyright Royalty Board (CRB), 294
copyright-free material, 258–259
CORE (Critical Operators and Radio Enthusiasts), 198–200
CORES system (Commission Registration System), 46–47, 62, 294
Corporation for Public Broadcasting (CPB), 294
Coursera, 189
covers, 116
Cox, James, 267
CP (Construction Permit), 61, 294
CRB (Copyright Royalty Board), 294
Creative Commons, 258
crew
  about, 197–198
  addressing staff responsibilities, 204–208
  attending staff meetings, 206
  clocking volunteer hours, 206
  covering scheduled shifts, 204–206
  cultivating respectful workplaces, 207–208
  developing station CORE, 198–200
  handling applications, 201
  overseeing orientation, 202
  recruiting and training on-air team, 200–204
  representing stations on social media, 207
  schedule for, 201
  suspension of shows, 208
  training for new recruits, 202–204
  vetting new recruits, 202

Critical Operators and Radio Enthusiasts (CORE), 198–200
crowdfunding, 129
C-Span, 259
CSRS (Call Sign Reservation System), 294
cue, 185, 294
customer relationship management (CRM) systems, 152

D

Davis, Al (engineer), 98
DCMA (Digital Millennium Copyright Act), 144–145, 172, 248
dead air, filling, 257–263
decibel (dB), 294
defamation of character, 171–172
demo, 238, 294
Denon DN-700CB, 105
Denon DN-790, 105
development director, 199
Dial, Davyne, 280
Digital Alert Systems (DASDEC III), 70
digital boards, 98
digital files, managing, 122–123
digital libraries, building, 121–125
Digital Millennium Copyright Act (DCMA), 144–145, 172, 248
digital transmission, licensing music and, 145–146
Diliberto, John, 245
direct licenses, 172
direct URLs, 79
direct-drive motors, 103
Directors and Officers (D&O) insurance, 33–34
disappearing (delayed) talent, 218
Discord, 144
distorted reporting, avoiding, 164–165
The Diva Hour, 177
diversity, of ownership, 22–23
DJ H.A.L., 122–123
DJ names, 180–181
DJ streams, establishing, 250–251
DJ training programs, organizing, 232–233
DocuSign, 194
Dolphin Radio, 286
Dome City Radio, 272
donor appreciation, 148–149
DonorPerfect, 152
Downtown Radio, 286
Driftless Turmoil, 178
drive time, 294
Dunifer, Stephen, 10, 12
duty of care, as a legal duty of board members, 24
duty of loyalty, as a legal duty of board members, 24
duty of obedience, as a legal duty of board members, 24
dynamic mics, 99

E

EAS (Emergency Alert System)
  defined, 295
  installing, 70–71
Echevarria, Antonion Carlos, 192
Echoes, 245
Edison, Thomas, 266
education statement, 47–48
Effective Radiated Power (ERP), 67, 295
"effects" (FX), 295
EIN (Employer Identification Number), applying for, 30–31
electrical tape, 116
Ellinger, Jim, 272
Elsherif, Mike, 178
emergencies
  reporting on, 238–239
  scheduling and, 205–206
Emergency Alert System (EAS)
  defined, 295
  installing, 70–71
Emergency Test Reporting System (ETRS), 71, 295
Employer Identification Number (EIN), applying for, 30–31
engineering study, 41–42, 52
ENIGMA group, 263

equal airtime, 190
equal opportunities, 168
equipment
checking, 185
failing, 218–219
selecting for studios, 97–107
sourcing housing for, 63
training on, 203
ERP (Effective Radiated Power), 67, 295
Erratic, Ricky, 190
E-Staff (Executive Staff), 198
ETRS (Emergency Test Reporting System), 71, 295
Event Management For Dummies (Capell), 156
events
breaking, 189–190
interviewing guests for, 228
launching, 155–156
events calendars, 193
evergreen, 295
ex parte rules, 52
Executive Staff (E-Staff), 198
exit agreements, 222
expenses. *See* station expenses
experimental radio programs, 179–180
experimenting, 263
The Exploded View, 177
Extreme Independent Radio, 286

F

Facility ID Number (FIN), 47, 62
Faculty Registration Number (FRN), 295
fader, 295
failing equipment, 218–219
Fairey, Shepard, 221
Fass, Bob, 270–271
FCC. *See* Federal Communications Commission (FCC)
Federal 501(c)3 tax-exempt organizations, 29, 30
Federal Communications Commission (FCC)
about, 8, 16, 19, 20, 161–162
Audio Division, 46
Daily Digest, 50
defined, 295
EAS Operating Handbook, 71
FCC Registration Number (FRN), 47
laws of, 160–171
Licensing and Management System, 63, 72, 73
registering with the, 46–47
self-inspection checklist, 72
website, 40, 46, 49, 71, 160
Federal Emergency Management Agency (FEMA), 70, 272
feed, 295
feedback, 295
festivals
interviewing guests for, 228
taking part in, 234–235
filing
annual taxes, 153
990s, 32–33
Film Fatale, 178
FIN (Facility ID Number), 47, 62
financial records, maintaining, 152–153
first adjacent channels, 41
First Amendment rights, 163, 164
1st adjacent channel, 293
501(c)3 status, establishing, 31–32
Floyd, George, 273
Fluxus movement, 182
Fly-In, 270
FM (Frequency Modulation)
about, 16–18
applying for, 46–49
defined, 295
finding, 39–42
FM transmitters, 66–67
interference, 216–217
licensing and FM transmission, 144–145
winning, 50–51
foley, 295
Form 211 Schedule 318 "Low Power Construction Permit," 49
Form 990-N (e-Postcard), 32

Form 990s, filing, 32–33
Form 1023 "Application for Recognition of Exemption under Section 501(c)3 of the Internal Revenue Code," 31
Form 1023-EZ, 31
format, 295
Forward Radio, 286
Fox, George, 232
Free Music Archive, 173, 259
Free Radio Berkeley, 10
Free Vermont Radio, 286
Freedman, Ken, 131, 221, 259
The Freedom Hour, 231
Freeform Portland KFFP-LP, 201, 286
Freesound, 173
Frequency Modulation (FM)
  about, 16–18
  applying for, 46–49
  defined, 295
  finding, 39–42
  FM transmitters, 66–67
  interference, 216–217
  licensing and FM transmission, 144–145
  winning, 50–51
Freshbooks, 152
Friends of WPVM, 28
FRN (Faculty Registration Number), 295
Frogtown Community Radio, 286
FundRazr, 129
funds
  for continuing education, 149–150
  hosting fundraising events, 138–140
  laws for fundraising, 170
  receiving, 34
furniture, for studios, 95–97
FX ("effects"), 295

G

gain, 295
Garage Band, 106
GatesAir, 66
general manager, 198–199
getting started
  about, 61
  choosing between on-site and remote-site transmission, 65–66
  claiming your call sign, 62–63
  FM transmitters, 66–67
  installing antennas, 68–69
  installing Emergency Alert System (EAS), 70–71
  keeping information safe, 62
  official launch, 74
  setting up your signal, 63–64
  turning it on, 71–73
GiveButter, 140
GoFundMe, 129
Goodnight Locker, 181
Google Voice, 34, 142
GoogleMeet, 144
Gorman-Redlich (EAS 1 + CAP-DEC 1), 70
government money, finding, 136
Grant, Clark, 279–280
Grant Writing For Dummies, 7th Edition (Browning), 136
graphic designer, 199
Grassroots Radio Conference (GRC), 150, 243–244, 295
Greb, Gordon, 267
GreenGlue, 95
Groves, Jake, 177
guests
  considerations for, 182
  interviewing, 227–230
Guitar Center, 102, 104
Guthrie, Arlo, 187

H

H.A.L., 295
ham radio club, 43
harassment, 208
Hardin, Warren, 267
Harvard Broadcasting Inc., 287
Haynes, Bill, 85
headlines, 295

headphone amps, 101–102
headphones, 101
Height Above Average Terrain (HAAT), 8, 67, 295
Herrold, Sybil, 266–267
Hill, Lewis, 131, 268–269
History San José, 267
hoaxes, avoiding, 164–165
Hoffman, Abbie, 270
holiday schedules, 205
Hollow Earth Radio, 287
homebases, maintaining, 142
hook, 296
hosts/hosting
  about, 280
  board meetings, 26
  fundraising events, 138–140
  on-air guests, 193–194
  open studio events, 226
  station spectacles, 236
  for web streams, 79
House of Representatives, 191
house rules, 173–174
Huckins, Dave, 138
Human Rights Radio, 10
Hundred Proof Fountain, 187
Hurricane Katrina, 272

I

icons, explained, 3–4
Illinois Valley Community Radio, 287
imminent lawless action, 165
incentive programs, 136–137
inclement weather, scheduling and, 205
income. *See* station income
incorporating nonprofit organizations, 29–30
indecent programming, 166–167
Indians of All Tribes (IAT), 271
Indiegogo, 129
informal objection, 51
in-house productions, archiving, 125–126
Inside a Question, 179
installing
  antennas, 68–69
  Emergency Alert System (EAS), 70–71
  studio furniture, 95–97
instruments, playing, 260–261
Integrated Public Alert & Warning System (IPAWS), 70
interference, FM, 216–217
interior spaces, 94–95
Internal Revenue Service (IRS), 31, 153
International Human Rights, 273
International Media Action, 21, 41, 64
Internet Archive, 259
Internet services, price of, 143
internships, 232–233
interviews, of guests, 227–230, 260
IT specialist, 199

J

Jampro JLLP-11, 68
Jazz ATX, 287
jingle, 296
Johnson, Carly, 177
Joppa Radio, 287

K

Kaatskit Radio Company, 98
KAAY, 185
Kale, LaGanzie, 275–276
Kantako, Mbanna, 10, 12
Kavoosi, Shadi, 231
KBFG, 287
KBMF Butte, 279–280, 287
KBOO, 114
KCIW, 287
KDIF 102.9 FM, 287
KDRU 98.1 - Drury University, 287
Kelley, Max, 279
Kennard, William, 11
KFJC Los Altos, 125
KFOK: The Voice of The Divide, 287

Kimel, Kris, 76–77
King, Martin Luther, Jr., 269–270
Klann, David, 124–125
KLEK Jonesboro, 275–276, 287
KLLG - Willits Hometown Radio, 287
KMGP Seattle, 276–277
KMRD Madrid, 28, 283–284
Knowledge, Understanding, & Wisdom, 287
Knoxville Community Radio, 287
Koch, Howard, 267–268
Kolehmainen, Mathias, 85
KOWS 92.5 FM, 287
KPFA, 268–269
KPPQ-LP Ventura, 288
KPSQ - The Public Square, 142, 288
KRWV, 288
KRYZ Radio, 288
KSFP - San Francisco Public Press, 288
KTQA, 288
KUHS, 288
KUOZ - University of the Ozarks, 288
kut, 296
KUYI Hopi Radio, 288
KVRU, 288
KWSI, 288
KXSU - Seattle University, 288
KZAX - Make Shift Radio, 288

L

Lacy, JohnBen, 178
Lakes Community Broadcasting, 288
landlords, negotiating with, 45–46
lawless action, inciting, 165
LCRA (Local Community Radio Act of 2010), 7, 11, 12–13, 20, 40, 296
legal advice, 160
legal ID, 168, 188, 296
legal identifiers, announcing, 188–189
letter of reasonable assurance, 48–49
level, 296
Lexington Community Radio/El Pulso Latio, 288
LibreTime, 124–125
LibriVox, 259
licensees, working with in your community, 55–56
licenses
  about, 144–146
  renewing, 72–73
Licensing and Management System (LMS), 49, 296
Listen2myradio, 85
listeners
  about, 241
  promoting internationally, 242–244
  providing on-demand listening, 248–250
  remote, 250–253
  syndicating programs worldwide, 245–248
Little Green Light, 152
The Little Ham Program, 266
live, going, 186–193
Live 365, 145
live read, 296
local businesses, interviewing guests for, 227
Local Community Radio Act of 2010 (LCRA), 7, 11, 12–13, 20, 40, 296
local events, covering, 236–238
local music director, 199
local performers, 261
locations, finding for studios, 92–94
log, 296
LogMeIn, 65, 143
logos, creating, 35
Long Beach Public Radio, 288
Louisville Visual Art (LVA), 93
Low Power FM (LPFM), 7–8, 296
Lumpen Radio, 288

M

Madame CJ Walker Museum, 270
Madrid Community Radio, 289
Mailchimp, 151
Manley, Brian, 178

Marconi, Guglielmo, 265–266
marking materials, 120
materials
for finding frequencies, 47–49
marking, 120
M-Audio Bx5 D3, 102
Maully, DJ, 192
MegaSeg Pro, 123, 124–125
membership, building for stations, 131–133
membership drive, 131
Memorandum Opinion and Order, 50–51
merchandise, 138, 154–155
Mercury Theater, 164
mic booms, 100
microphones, 99–100, 262
Midnight Locker, 181
Mightycause, 129
Mission Statement, defining your, 27–28
mix, 296
Mixcloud, 173, 249
mixer, 98–99
The Mix 94.3, 289
mock inspection, 150
Money-O-Meter, 129
monitors, 102, 296
mono, 296
Morris, Tee (author)
Podcasting For Dummies, 250
Mueller, John Paul (author)
Android Application Development All-in-One For Dummies, 85
multi-colored electrical tape, 116
Murphy's law of radio
about, 211–212
breaking rules, 219–221
carrying on, 221–223
disappearing (delayed) talent, 218
failing equipment, 218–219
FM interference, 216–217
preparing for silent broadcasts, 212–215
radio silence, 214
spilling onto soundboard, 219
stream glitching, 217–218
streaming silence, 214–215
user error, 213
utility failure, 213
music
about, 111
albums, 118–120
archiving original in-house productions, 125–126
building digital libraries, 121–125
building physical libraries, 112–120
cassette tapes, 118
CDs, 117–118
choosing organizational methods for, 115
collecting materials for, 116
cueing, 185
for filling dead air, 257–258
licensing for, 144–146, 172–173
marking materials, 120
selecting, 181–182
shelving for, 113–114
vinyl, 116–117
music director, 199
Music Go Round, 104
music programs, 177–178
mutually exclusive group (MX group), 51–55, 296
My Zone, 138

N

naming nonprofit organizations, 27
Napa Radio Project, 289
NASA, 259
National Association of Black Owned Broadcasters (NABOB), 244
National Association of Broadcasters (NAB), 11, 151, 244, 296
National Association of Hispanic Journalists (NAHJ), 244
National Association of Music Merchants (NAAM), 102
National Federation of Broadcasters (NFCB), 150

National Federation of Community Broadcasters (NFCB), 102, 244, 296
National Public Radio (NPR), 11, 187, 296
Nautel, 66
NavePoint, 63
NCE (NonCommercial Educational) broadcast station, 296
needle, 218
Neon, 152
news, reading, 189
news director, 199
NiCOM, 66, 68
Nielsen radio markets, 20
95.7 The Boss, 285
99.1 The Beat, 285
99.3FM WBTV-LP, 285
Nobex, 85
Noguchi, Shurtaro, 181
NonCommercial Educational (NCE) broadcast station, 296
noncommercial stations, 169–170
non-disclosure, 222
nonprofit organizations
  about, 19
  application for Low Power FM Construction Permit, 20–23
  branding, 34–37
  forming new, 24–34
  incorporating, 29–30
  interviewing guests for, 227
  naming, 27
  partnering with existing, 23
nonprofit postage, 33
NPR (National Public Radio), 11, 187, 296
numbers station, 262–263

O

O. C. White, 100
objectionable programming, 165–167
obscene programming, 166–167
Occam Design, 133
official launch, 74
official station identification, 168
Ohm Radio, 289
on-air guests, hosting, 193–194
on-air light, 100
on-air team, recruiting and training, 200–204
on-demand listening, providing, 248–250
101 Jamz, 285
101.9 WKRP, 286
1013krd Rockford's Best Mix, 286
OneCause, 140
online radio directories, registering with, 84–85
on-site transmission, remote-site transmission compared with, 65–66
open studio events, hosting, 226
opportunities, creating, 128
organizational methods, choosing for music collection, 115
orientation, overseeing, 202
oversight board, 24–25
ownership, diversity of, 22–23

P

Pacifica Affiliates Network, 79, 150–151, 173, 245, 268–269, 297
Pacifica Archives, 272
Pacifica Foundation, 297
Pacifica's Audioport, 245
Pacifica's Confessor, 146, 172, 249
The Panic Broadcast, 267
parades, taking part in, 234–235
Part 15 transmitter, 56–57
partnerships
  developing, 133–134
  with nonprofits, 23
Party Points, 192
payola, 171
PayPal, 129
peace clock, 188
People of Color Caucus, 244

Performing Rights Organizations (PROs), 144, 172, 297
personnel, hiring, 147–150
petition to deny, 51
Phillips, Brice, 272
Phillips, Damaris, 179
PhillyCAM Radio, 289
physical libraries, building, 112–120
Pirate March, 11
pirate radio movement, 9–12, 52
Pirate ship, 279
player, 106
player buttons, embedding on websites, 80–81
playlist, 297
Podcasting For Dummies (Morris and Tomasi), 250
Poder 97.1 FM, 289
political broadcasts, balancing, 165
politicians, interviewing guests for, 228
pop-up players, 81
Portland Radio Project, 289
president, of boards, 26
Presidential elections, 267
Priority Points, 22, 23, 54–55
profane programming, 166–167
professional broadcaster's policy, 34
professional organizations, joining, 150–151
program director, 199
program titles, 180–181
programmer, 297
programming
  about, 175–177
  albums, 118–120
  announcing legal identifiers, 188–190
  being "in the now," 186–187
  checking equipment, 185
  committee, 201
  cueing music, 185
  DJ names, 180–181
  events calendars, 193
  experimental radio programs, 179–180
  going live, 186–193
  guests for shows, 182
  handling surprise situations, 187–188
  hosting on-air guests, 193–194
  making music selections, 181–182
  making public service announcements, 191–192
  music programs, 177–178
  preparing shows, 184–185
  program titles, 180–181
  promoting your program, 183
  reading out promos, 192
  reading the news, 189–190
  setting the scene, 180–184, 187
  spoken word programs, 178–179
  studio etiquette, 194–196
program-related materials, announcing, 170–171
Prometheus Radio Project, 11, 12, 41, 64, 68, 160
promos
  defined, 297
  reading out, 192
promoting
  radio programs, 183
  stations internationally, 242–244
promotions director, 199
PROs (Performing Rights Organizations), 144, 172, 297
PS Form 3624 "Application to Mail at Nonprofit USPS Marketing Mail Prices," 33
PSA director, 199
Public Address (PA) system, 237, 297
public affairs, 190
Public Radio eXchange (PRX), 173, 245–246, 297
public safety organizations, 23
Public Service Announcement (PSA)
  defined, 297
  making, 191–192
publication, for promoting internationally, 243
The Public and Broadcasting, 162, 164, 165, 166–167, 169
The Public Square, 289

Q

Q - 101.3, 289
QuickBooks, 152
QuietRock, 95

R

Radio Act of 1914, 9
Radio Alcatraz, 271–272
Radio Data System (RDS), 67, 146, 297
radio directories, registering with, 84–85
Radio Garden, 84, 85
Radio Mutiny, 11
Radio Preservation Act, 11
radio programs
  preparing, 184–185
  promoting, 183
  recommended, 265–273
radio silence, 214
radio stations
  about, 13–15
  directory of, 285–291
Radio Survivor, 97, 243
Radio Tacoma, 289
Radio Unamable, 270–271
Radio World, 243
RadioDJ, 124–125
RadiOpio - Pa'ia Youth and Cultural Center, 289
Radius Maps app, 281
rallies, 238
Ramsey Electronics, 66
RDS (Radio Data System), 67, 146, 297
Real Oldies 97.9 The WREN, 289
REC Networks, 20, 41, 167
recruiting on-air team, 200–204
Red Cross Emergency app, 239
reengineering escape, 51
Reeves, Martha, 146
registering
  with the FCC, 46–47
  with online radio directories, 84–85
release, 297
release forms, 229
Remember icon, 3
remote, 297
remote access, price of, 143
remote set-ups, 250–253
remote-site transmission, on-site transmission compared with, 65–66
renewing licenses, 72–73
Required Monthly Tests (RMT), 71, 297
reverse shadowing, 203
Richardson, Kara, 281–282
Richmond Independent Radio, 289
Rivendell, 124–125
Riverside, 251
Riverwest Radio, 289
Rocket Broadcaster, 81
RockHits 92.3, 289
Rocky Mountain Multicultural Community Radio, 289
Ross, Joan C., 282–283
Ross, Julianna, 276–277
rotation, 297
royalties, 297
Ruedenberg, Ursula, 246–248
rules and responsibilities
  about, 159–160
  announcing program-related materials, 170–171
  content regulations, 167–169
  defamation of character, 171–172
  distorted reporting, 164–165
  FCC laws, 160–171
  finding legal advice, 160
  First Amendment rights, 163
  fundraising, 170
  hoaxes, 164–165
  house rules, 173–174
  music licensing, 172–173
  objectionable programming, 165–167
  payola, 171
  political broadcasts, 165

The Public and Broadcasting, 162
rule-breaking, 219–221
rules and regulations training, 203
staying noncommercial, 169–170

S

Sabzevari, Nilou, 231
Sage Alert Systems (Endec 3644), 70
SBE (Society of Broadcast Engineers), 297
scenes, setting, 180–184
schedule, for on-air team, 201
scheduled shifts, covering, 204–206
2nd adjacent channel, 41, 293
secretary, of boards, 27
segue, 297
Seiller Waterman, 133
Selby, Max, 233–234
Selby, Sean, 96
Sendinblue, 151
Sennheiser MD 421-IIs, 99
SESAC (Society for European Stage Authors and Composers), 144–145, 298
Seven Second Delay, 221
shadowing, 203
sharpies, 116
shelf divider cards, 116
shelving, for music collection, 113–114
Shotgun!, 181
Shure SM57, 100
sickness, scheduling and, 205–206
Sieg, Wilfred, 93
signal, setting up your, 63–64
silent auction, 140
Simon, Paul, 112
Simplecast Creator Connect, 250
simulcasting, 261–262, 273, 298
singleton, 298
Site To Location (STL) link, 65
skills, for smart speakers, 88
smart speakers, skills for, 88
Smooth Jazz Utah, 289
social media
establishing a presence on, 36
for promoting internationally, 242–243
representing stations on, 207
Society for European Stage Authors and Composers (SESAC), 144–145, 298
Society for HR Management, 208
Society of Broadcast Engineers (SBE), 297
software, using to play music, 123–125
sound dampening, 95
sound drop, 298
soundbite, 298
soundboard, 98–99, 219
Soundcloud, 249
SoundExchange, 145–146, 298
SoundExchange Licensee Direct, 146
SoundForge, 106
soundproofing, 95
Soundstack, 79
sources
audio, 173
equipment housing, 63
Space 101.1 FM, 289
Spark!, 290
Special Temporary Authority (STA), 73, 298
Spinitron, 146, 172, 248, 249, 298
spoken word programs, 178–179
sponsorships
developing, 133–134
identification of, 169
STA (Special Temporary Authority), 73, 298
The stacks, 298
staff appreciation, 148–149
staff meetings, attending, 206
staff responsibilities, 204–208
StarTech, 63
State Emergency Alert Plan, 71
State Emergency Communications Committees, 71
state sales tax exemption, 32
station archivist, 199
station authorization, 298

station expenses
about, 141
buying merchandise, 154–155
costing for communications, 151
factoring in price of technology, 143–144
funding continuing education, 149–150
hiring personnel, 147–150
joining professional organizations, 150–151
launching events, 155–156
licensing music, 144–146
maintaining financial records, 152–153
maintaining homebase, 142
managing, 141–156
showing staff/donor appreciation, 148–149
station updates, 156
sundries, 154
station founders, 275–284
station income
about, 127
building station membership, 131–133
community giving days, 130–131
creating opportunities, 128
developing, 127–140
developing partnerships/sponsorships, 133–134
finding government money, 136
hosting fundraising events, 138–140
incentive programs, 136–137
making friends, 128
merchandise sales, 138
organizing capital campaigns, 128–131
vehicle donations, 137
working with underwriters, 134–135
station interns, 199
station membership, building, 131–133
Station Playlist, 124–125
station spectacles, hosting, 236
station updates, 156
station-conducted contests, 168
statutory licenses, 144, 172
stereo, 298
stereophonic sound, 298
Stitcher, 250
STL (Site To Location) link, 65
Stockwell, Norm, 77
stream glitching, 217–218
Streamguys, 79
streaming silence, 214–215
Streema, 84
Strummer, Joe, 112–113
studio manager, 199
studio monitors, 102
Studio to Transmitter Link (STL), 298
studios
about, 91
etiquette for, 194–196
finding locations for, 92–94
installing studio furniture, 95–97
interior spaces, 94–95
providing community access and security, 108
selecting equipment for, 97–107
stylus, 218
subbing, 204
Subsonic, 123
sundries, 154
Surf 97.3 FM, 290
surprise situations, handling, 187–188
suspension, of shows, 208
swag, for promoting internationally, 242
Swain, Lee Jaxon, 192
Sweetwater, 97, 100
Swift, 85
syndicated programs, 245–248, 298

T

The Takeaway, 245
Takoma Radio, 290
Tascam CD200s, 105
Taylor, Breonna, 190, 273
Teamviewer, 65
teaser, 298
Technical Stuff icon, 3
Technics SL-1200s, 103
technology, price of, 143–144

Teen Radio Academy, 233
Telecommunications Act of 1996, 11
telephone conversations, broadcast of, 168
telephone-hybrid, 260
tentative selectee, 51
Tesla, Nikola, 9, 265–266
32auctions, 140
Tieline, 65
timeshares, joining, 54–55
Tip icon, 4
Tomasi, Chuck (author)
  Podcasting For Dummies, 250
Total Power Output (TPO), 67, 298
towers, finding, 42–46
Tracey, Rage C. (author)
  Accounting For Dummies, 7th Edition, 152
Tracy, John A. (author)
  Accounting For Dummies, 7th Edition, 152
training, for on-air team, 200–204
training director, 199
transition, 298
transmission, on-site compared with remote-site, 65–66
transparency, maintaining, 153
treasurer, of boards, 27
tribal applicants, 22
Tridish, Pete, 12
Tripp Lite, 63
tropospheric ducting, 216
Trudell, John, 271–272
Tune-In, 84
turntables, 102–103

U

underwriting, 134–135, 169–170
Uninterruptible Power Supply (UPC), 107
Urick, Todd, 44, 53–54, 68
URLs, direct, 79
U.S. Copyright Office, 145, 258
U.S. Equal Employment Opportunity Commission (EEOC), 207–208
US National Electrical Code (NEC), 69
user error, 213
USPS (website), 33
utility failure, 213

V

vehicle donations, 137
Venmo, 129
Veterans History Project (VHP), 231
veterans' stories, 231–232
VIAVI (EASyCAP B3030), 70
vice president, of boards, 27
video conferencing, price of, 144
vinyl, 116–117
vinyl letters, 116
Virtual GRC, 192
Vision Statement, creating, 28
voice, amplifying, 259–260
The Voice of Arkansas Minority Advocacy Council/KLEK Jonesboro, 28
The Voice of Athol-Orange, 290
The VOICE of the Community, 290
volunteer hours, 206
vox pop, 237
VU meter, 298

W

W3 Webschools, 81
Waits, Jennifer, 97
WAKT, 290
WAM (Women's Audio Mission), 244, 298
War of the Worlds, 267–268
Warning icon, 4
Wave Farm Transmission Art Installation Park, 179, 263
WAYO Rochester, 279, 290
WBCA - Boston Neighborhood Network, 290
WCAA LP 107.3 FM, 290
WCFA - Cape May Radio, 290
WCXP Chicago, 277–278

WDZD - Fun One Radio, 290
Weather Warlock, 180
web designer, 199
web player icon, 80
web stream
about, 75–78
adding audio, 81–83
connecting with hosts, 79
developing apps, 85–88
direct URLs, 79
embedding player buttons on website, 80–81
launching, 78–83
registering with online radio directories, 84–85
skills for smart speakers, 88
utilizing pop-up players, 81
webcaster, 105
webhosting, price of, 143
websites
building, 36
embedding player buttons on, 80–81
numbers stations, 263
WEFT, 114
Welles, Orson, 267
WELT, 290
WERA - Arlington Independent Media, 290
WERD, 269–270
WFHB Bloomington, 125
WGRN - The Green Renaissance, 290
WGXC 90.7 Arca, 263
Wheatstone, 98
WHIV, 290
Wi-Fi hotspots, 213
Wilkin, Claudia Ann, 10
Wimberley Valley Radio, 112, 290
Wix, 148
WNUC Detroit, 282–283, 291
WNUW - Newman University, 291
Women's Audio Mission (WAM), 244, 298
Wood, Darrick, 179
WordPress, 148
working board, 24–25
workplaces, cultivating respectful, 207–208
worldwide syndication, 245–248
WPPM 106.5 FM, 291
WPVM Asheville, 280, 291
WQRT, 291
WQRZ-LP, 272
WRGV - Georgetown College, 291
WSLR Sarasota, 282, 291
WTSQ: The Status Quo, 291
WURK Community Radio, 291
WXNA, 47–48, 291
WXND Louisville, 281–282
WXRU-LP SMOOTH 107.9 FM, 291
WXTJ: University of Virginia, 291

X

Xerocraft, 291
XLR cables, 237–238
Xtendsound, 291

Y

Yellowtech, 100
youth groups, 231

Z

Zoho Books, 152
zoning, 52
ZUMIX Radio, 291

About the Author

Sharon Scott is the co-founder and general manager of ART FM / WXOX 97.1 FM Louisville, a noncommercial radio station committed to providing artists and community members with access to the airwaves for creative and experimental use. Built with her husband Sean and their son Max, WXOX currently has over 100 volunteer DJs bringing entertainment and new ideas to listeners all over the world. In addition to the local FM broadcast, the ART FM stream is available worldwide through the website `artxfm.com` and free WXOX apps.

Born in Atlanta, Georgia, Sharon Scott grew up on WREK 91.1 FM and began her radio career at WRVU 91.1 FM Nashville, "The Voice of Vanderbilt University." After graduating, Scott and her colleagues established WRVU Friends & Family, the nonprofit organization that now operates WXNA on 101.5 FM Nashville.

Scott is the founder of LPFM Solidarity and the Virtual GRC nationwide networks of community radio leaders, and she serves on the board of the Association of Affiliates (AOA) representing over 200 Pacifica-affiliated stations. Sharon is dedicated to uniting community radio stations nationwide and hopes that this book will be a source for building nonprofit networks. After the passage of the *Local Community Radio Act* and the global proliferation of listening online, she sees a bright future for creative broadcasting and a powerful force in community radio united.

Dedication

To the Pirates who paved the way.

To Sean and Max who helped see it through.

Author's Acknowledgments

This book, like my radio dreams, would not exist if it weren't for my partner-in-crime Sean Selby. Nor would my heart sing so brightly were it not for our son Maxwell Cash. Love you both. You are my world.

To my knowledge this is the only LPFM handbook of its kind, and I thank Myka Carroll for recognizing the need to get this information in print. Great appreciation goes to Wiley executive editor Steve Hayes for representing the publication and wrangling it into existence. Sincere appreciation to my editor Daniel Mersey for a keen and subtle twist of pen. Thank you also to Vicki Adang and Kristie Pyles.

I am grateful to everyone at Wiley who has provided me with the opportunity to write this book. It is an honor to put this all into print.

The LPFM movement has everything to do with community, so it would not make any sense for me to write this book alone. Thank you to all the radio colleagues who have contributed throughout. David Klann of Broadcast Tool & Die served as the technical advisor to this manuscript and offered fair and balanced advice. Kyle Anne Citrynell of Seiller Waterman LLC poured her ever-valuable insight into Chapter 2, sharing the wisdom that got ART FM started. Todd Urick, chief technical officer of Common Frequency was generous, as always, with the engineering information he provided for Chapters 3 and 4. I am deeply appreciative of you all and your unending dedication to community radio.

Thank you always to my mother Nancy B. Scott and her partner George W. Fox. I have been blessed to have their input and encouragement along the way. Thank you to my father Ken Scott; I am proud to have his sense of adventure that would never allow a project of this magnitude to pass me by. Thank you to the children Naomi, Lennon, Blessing, and Stevie-Grace for always introducing me to new ideas and keeping me on my toes!

Thank you to John Begley, Margue Esrock, Kris Kimel, Leslie Millar, and Anna Tataman, and to all those who encountered ART FM as an idea and believed in it enough to make it real. Thank you to the entire XOX family. You have entertained, delighted, surprised, and impressed me since day one — but never more-so than now. This manuscript has been a time-consuming process, but it has been a pure delight to hear you all keeping the station live and legit even when I've had to lock myself away from all the excitement. You are the ultimate success. Now it's time to party!

XOX,

DJ JET PILOT

Publisher's Acknowledgments

Acquisitions Editor: Steve Hayes
Project Manager: Dan Mersey
Technical Editor: David Klann
Senior Managing Editor: Kristie Pyles
Production Editor: Tamilmani Varadharaj
Cover Image: © avdyachenko/Getty Images

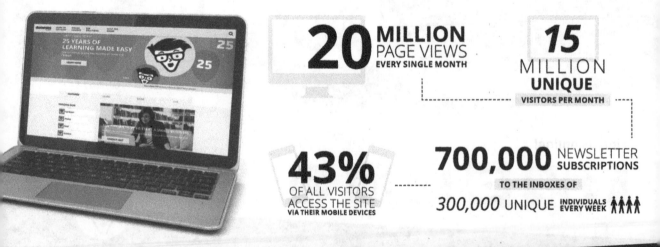

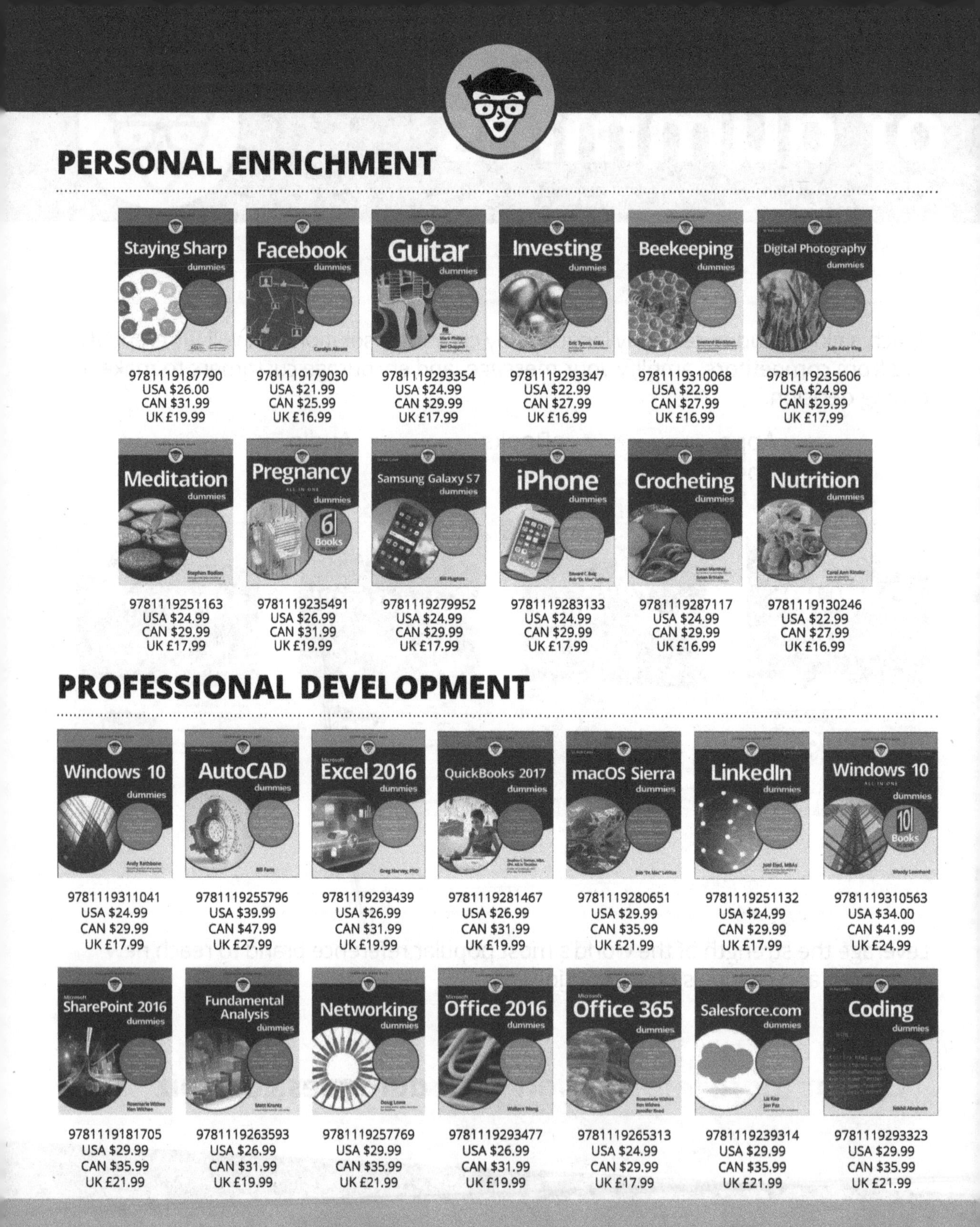

PERSONAL ENRICHMENT
Staying Sharp dummies
9781119187790
USA $26.00
CAN $31.99
UK £19.99
Facebook dummies
9781119179030
USA $21.99
CAN $25.99
UK £16.99
Guitar dummies
9781119293354
USA $24.99
CAN $29.99
UK £17.99
Investing dummies
9781119293347
USA $22.99
CAN $27.99
UK £16.99
Beekeeping dummies
9781119310068
USA $22.99
CAN $27.99
UK £16.99
Digital Photography dummies
9781119235606
USA $24.99
CAN $29.99
UK £17.99
Meditation dummies
9781119251163
USA $24.99
CAN $29.99
UK £17.99
Pregnancy dummies
6 Books
9781119235491
USA $26.99
CAN $31.99
UK £19.99
Samsung Galaxy S7 dummies
9781119279952
USA $24.99
CAN $29.99
UK £17.99
iPhone dummies
9781119283133
USA $24.99
CAN $29.99
UK £17.99
Crocheting dummies
9781119287117
USA $24.99
CAN $29.99
UK £16.99
Nutrition dummies
9781119130246
USA $22.99
CAN $27.99
UK £16.99
PROFESSIONAL DEVELOPMENT
Windows 10 dummies
9781119311041
USA $24.99
CAN $29.99
UK £17.99
AutoCAD dummies
9781119255796
USA $39.99
CAN $47.99
UK £27.99
Excel 2016 dummies
9781119293439
USA $26.99
CAN $31.99
UK £19.99
QuickBooks 2017 dummies
9781119281467
USA $26.99
CAN $31.99
UK £19.99
macOS Sierra dummies
9781119280651
USA $29.99
CAN $35.99
UK £21.99
LinkedIn dummies
9781119251132
USA $24.99
CAN $29.99
UK £17.99
Windows 10 dummies
10 Books
9781119310563
USA $34.00
CAN $41.99
UK £24.99
SharePoint 2016 dummies
9781119181705
USA $29.99
CAN $35.99
UK £21.99
Fundamental Analysis dummies
9781119263593
USA $26.99
CAN $31.99
UK £19.99
Networking dummies
9781119257769
USA $29.99
CAN $35.99
UK £21.99
Office 2016 dummies
9781119293477
USA $26.99
CAN $31.99
UK £19.99
Office 365 dummies
9781119265313
USA $24.99
CAN $29.99
UK £17.99
Salesforce.com dummies
9781119239314
USA $29.99
CAN $35.99
UK £21.99
Coding dummies
9781119293323
USA $29.99
CAN $35.99
UK £21.99